TABLE OF CONTENTS

Geometry for Primar[y]

Introduction 2	
Letter to Parents 4	
Letter to Students 5	
Student Progress Chart................................ 6	
NCTM Standards Correlation 8	

UNIT 1: 2-Dimensional Figures
- Assessment 9
- Sides and Corners 10
- Sides and Corners 11
- Line Segments 12
- Sides, Corners, and Line Segments 13
- Sides, Corners, and Line Segments 14
- Sides and Corners 15
- Plane Figures 16
- Plane Figures 17
- Plane Figures 18
- Venn Diagram 19
- Identifying Plane Figures 20
- Identifying Plane Figures 21
- Attributes of Plane Figures 22
- Drawing Plane Figures 23
- Drawing Plane Figures 24
- Drawing Plane Figures 25
- Making a Table 26
- Problem Solving 27
- Problem Solving 28

UNIT 2: 3-Dimensional Figures
- Assessment 29
- Solids 30
- Solids 31
- Solids 32
- Solids and Plane Figures 33
- Solids and Plane Figures 34
- Identifying Solids 35
- Identifying Shapes 36
- Attributes of Solids 37
- Attributes of Solids 38
- Problem Solving 39

UNIT 3: Congruence
- Assessment 40
- Matching Shapes 41
- Same Shape, Same Size 42
- Same Shape, Same Size 43
- Drawing Congruent Shapes 44
- Drawing Congruent Figures 45
- Problem Solving 46

UNIT 4: Sy[mmetry]
- Assessment ....[48]
- Identifying [Symmetry] ....[ ]
- Identifying Symmetry 49
- Drawing Lines of Symmetry 50
- Drawing Lines of Symmetry 51
- Problem Solving 52

UNIT 5: Perimeter and Area
- Assessment 53
- Measuring Perimeter 54
- Measuring Perimeter 55
- Measuring Perimeter 56
- Measuring Perimeter 57
- Perimeter/Area 58
- Perimeter/Area 59
- Perimeter/Area 60
- Problem Solving 61
- Problem Solving 62
- Perimeter Using Scale.............................63
- Estimating Perimeter 64
- Using a Calculator to Find Perimeter 65
- Area 66
- Area 67
- Problem Solving 68

UNIT 6: Fractions Using Pictorial Models
- Assessment 69
- Meaning of Fractions 70
- Meaning of Fractions 71
- Meaning of Fractions 72
- Identify Equal Parts 73
- Equal Parts 74
- Equal Parts 75
- Fractions 76
- Fractions 77
- Writing Fractions 78
- Writing Fractions 79
- Writing Fractions 80
- Writing Fractions 81
- Writing Fractions 82
- Identifying Fractional Parts 83
- Identifying Fractional Parts 84
- Identifying Fractional Parts 85
- Problem Solving 86

UNIT 7: Patterns and Coordinate Graphs
- Assessment 87
- Completing a Pattern 88
- Completing a Pattern 89
- Completing a Pattern 90
- Problem Solving 91
- Using a Coordinate Graph 92
- Using a Coordinate Graph 93
- Problem Solving 94

ANSWER KEY..95

INTRODUCTION
Geometry for Primary Grade 2

Helping students form an understanding of geometric shapes is a challenging task. In order to help students learn to recognize shapes, they must be approached in terms that will have meaning for them. The National Council of Teachers of Mathematics (NCTM) has set specific standards to help students become confident in their mathematical abilities. Geometry is an important component of the primary mathematics curriculum because geometric knowledge, relationships, and insights are useful in everyday situations and are connected to other mathematical topics and school subjects. *Geometry for Primary Grade 2* blends the vision of the NCTM Standards.

Geometry for Primary Grade 2 provides you with the opportunity to expand students' knowledge of geometry. Activities requiring the identification of sides, corners, line segments, and congruent figures are presented. Students recognize the shapes and names for solid and plane figures. They also identify the number of sides and corners of squares, rectangles, and triangles. Recognizing square corners and line segments, identifying same size and same shape, and detecting symmetric figures and lines of symmetry are skills included in this book. Students continue their study by finding perimeter and area and learning the meaning of fractions and fractional parts of a whole. Students continue the study of pattern sequencing and working with a coordinate graph.

Art activities that involve symmetry will enhance the learning experience for most students, as will the discussion of shapes in the classroom or a scavenger hunt for shapes through the pages of magazines. Providing experiences for exploration of straight and curved lines in the classroom will make textbook pages come to life.

It is essential that students be given sufficient concrete examples of geometric concepts. Manipulatives that can be used to reinforce the skills are recommended on the activity pages.

Organization

Seven units cover the basic geometric skills presented in the second grade: 2-dimensional figures, 3-dimensional figures, congruence, symmetry, perimeter and area, fractions using pictorial models, and patterns and coordinate graphs. In *Geometry for Primary Grade 2* the mathematics curriculum is presented so that students can:

- formulate and solve problems from everyday and mathematical situations
- describe, model, draw, and classify shapes
- investigate and predict the results of combining, subdividing, and changing shapes
- develop spatial sense
- relate geometric ideas to number and measurement ideas
- recognize and appreciate geometry in their world
- understand the attributes of length, perimeter, and area
- develop concepts of fractions
- recognize, describe, extend, and create a variety of patterns

INTRODUCTION

Geometry for Primary Grade 2

   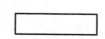

Use

The activities in this book are designed for independent use by students who have had instruction in the specific skills covered in the lessons. Copies of the activity sheets can be given to individuals or pairs of students for completion. When students are familiar with the content of the worksheets, they can be assigned as homework.

To begin, determine the implementation that fits your students' needs and your classroom structure. The following plan suggests a format for this implementation.

1. **Administer** the Assessment test to establish baseline information on each student. These tests may also be used as post-tests when students have completed a unit.

2. **Explain** the purpose of the worksheets to the class.

3. **Review** the mechanics of how you want students to work with the activities. Do you want them to work in pairs? Are the activities for homework?

4. **Introduce** students to the process and purpose of the activities. Work with students when they have difficulty. Give them only a few pages at a time to avoid pressure.

Additional Notes

1. <u>Parent Communication</u>. Send the Letter to Parents home with students.

2. <u>Student Communication</u>. Encourage students to share the Letter to Students with their parents.

3. <u>Manipulatives</u>. Manipulatives are recommended at the bottom of the activity pages. This optional feature can help you provide concrete examples to reinforce geometric concepts.

4. <u>NCTM Standards Correlation</u>. This chart helps you with your lesson planning. An icon for each standard is included on the student page so that you can tell at a glance what skill is being reinforced on the page.

5. <u>Student Progress Chart</u>. Duplicate the grid sheets found on pages 6-7. Record student names in the left column. Note date of completion of each lesson for each student.

6. <u>Have fun!</u> Working with these activities can be fun as well as meaningful for you and your students.

Dear Parent:

During this school year, our class will be working with mathematical skills. We will be completing activity sheets that provide enrichment in the area of geometry. This includes skills in problem solving, geometry and spatial sense, fractions, patterns, and coordinate graphs.

From time to time, I may send home activity sheets. To best help your child, please consider the following suggestions:

- Provide a quiet place to work.
- Go over the directions together.
- Encourage your child to do his or her best.
- Check the lesson when it is complete.
- Go over your child's work, and note improvements as well as problems.

Help your child maintain a positive attitude about mathematics. Let your child know that each lesson provides an opportunity to have fun and to learn. If your child expresses anxiety about these strategies, help him or her understand what causes the stress. Then talk about ways to eliminate math anxiety.

Above all, enjoy this time you spend with your child. He or she will feel your support, and skills will improve with each activity completed.

Thank you for your help!

Cordially,

Dear Student:

This year you will be working in many areas in mathematics. The activities in this program concentrate on the area of geometry. You will work with lines, sides, and corners; plane and solid figures; matching shapes that are the same shape and size; lines of symmetry; measuring the distance around and the area of figures; equal parts of a figure; patterns; and using a graph to find spots on a map. You will get to color, draw, count, and sort shapes, measure size, draw patterns, and solve problems. These activities will show you fun ways to practice geometry!

As you complete the worksheets, remember the following:

- Read the directions carefully.
- Read each question carefully.
- Check your answers after you complete the activity.

You will learn many ways to solve math problems. Have fun as you develop these skills!

Sincerely,

STUDENT PROGRESS CHART

STUDENT NAME	UNIT 1 2-DIMENSIONAL FIGURES																				UNIT 2 3-DIMENSIONAL FIGURES											UNIT 3 CONGRUENCE							UNIT 4 SYMMETRY				
	10	11	12	13	14	15	16	17	18	19	20	21	22	23	24	25	26	27	28	29	30	31	32	33	34	35	36	37	38	39	40	41	42	43	44	45	46	47	48	49	50	51	52

STUDENT PROGRESS CHART

STUDENT NAME	UNIT 5 PERIMETER AND AREA															UNIT 6 FRACTIONS USING PICTORIAL MODELS																UNIT 7 PATTERNS AND COORDINATE GRAPHS							
	54	55	56	57	58	59	60	61	62	63	64	65	66	67	68	70	71	72	73	74	75	76	77	78	79	80	81	82	83	84	85	86	88	89	90	91	92	93	94

www.svschoolsupply.com
© Steck-Vaughn Company

Geometry 2, SV 5806-X

NCTM STANDARDS CORRELATION

NCTM Standard	Unit 1	Unit 2	Unit 3	Unit 4	Unit 5	Unit 6	Unit 7
1: Problem Solving • formulate problems from everyday and mathematical situations	26, 27, 28	37, 39	46	51, 52	61, 62, 63, 68	75, 82, 83, 84, 85, 86	93, 94
9: Geometry & Spatial Sense • describe shapes	10, 11, 12, 13, 14, 15, 16, 22, 25, 27	36, 37, 38, 39					
9: Geometry & Spatial Sense • model shapes		30, 31, 32					
9: Geometry & Spatial Sense • draw shapes	12, 13, 14, 15, 23, 24, 25		42, 43, 44, 45, 46				88, 89, 90
9: Geometry & Spatial Sense • classify shapes	17, 18, 20, 21	30, 31, 32, 33, 34, 35	41, 42, 43				
9: Geometry & Spatial Sense • combining, subdividing, & changing shapes		37		48, 49, 50, 51, 52		70, 71, 72, 73, 74, 75, 76, 77, 78, 79, 80, 81, 82, 83, 84, 85, 86	
9: Geometry & Spatial Sense • develop spatial sense		31, 32, 33, 34, 36, 37, 38, 39		48, 49, 50, 51		92, 93, 94	
9: Geometry & Spatial Sense • relate geometric ideas to number and measurement ideas	18, 19, 20, 21, 26	35			54, 55, 56, 57, 58, 59, 60, 61, 62, 63, 64, 65		
10: Measurement • understand the attributes of length					54, 55, 56, 57, 58, 59, 60, 61, 62, 63, 64, 65		
10: Measurement • understand the attributes of area					59, 60, 66, 67, 68		
12: Fractions & Decimals • develop concepts of fractions						70, 71, 72, 73, 74, 75, 76, 77, 78, 79, 80, 81, 82, 83, 84, 85, 86	
13: Patterns & Relationships • recognize, describe, extend, and create a variety of patterns							88, 89, 90, 91

www.svschoolsupply.com
© Steck-Vaughn Company

Geometry 2, SV 5806-X

Name _____ Date _____

2-DIMENSIONAL FIGURES
Assessment: Unit 1

1. Color inside the circles blue.
   Color inside the triangles orange.
   Color inside the rectangles red.

   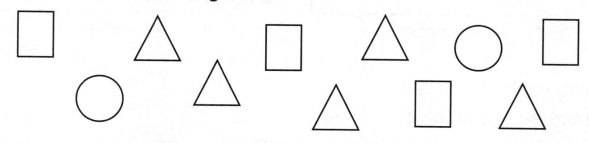

2. Count each shape. Write how many.

   _____ circles   _____ triangles   _____ rectangles

3. Color inside the triangles purple.
   Color inside the circles red.
   Color inside the squares green.

4. How many are there? Count each shape. Write how many.

   _____ triangles   _____ circles   _____ squares

5. Draw a shape that has 3 sides and 3 corners.

   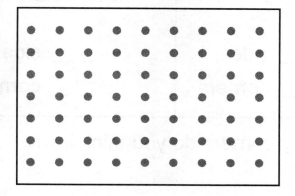

www.svschoolsupply.com
© Steck-Vaughn Company

9

Unit 1: 2-Dimensional Figures
Geometry 2, SV 5806-X

Name_____ Date_____

2-DIMENSIONAL FIGURES
Sides and Corners

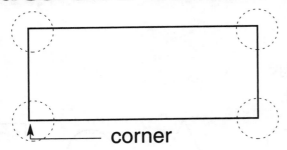 ← corner

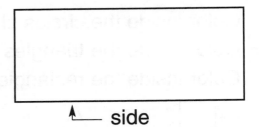 ← side

Ring each corner red.
Trace each side blue.
Then write how many sides and corners.

1.
   __4__ sides
   __4__ corners

2.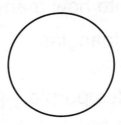
   _____ sides
   _____ corners

3.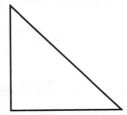
   _____ sides
   _____ corners

4.
   _____ sides
   _____ corners

5.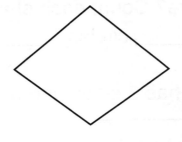
   _____ sides
   _____ corners

6.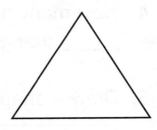
   _____ sides
   _____ corners

7. If a shape has 6 sides, how many corners do you think it will have? _____ corners

Sort Attribute Blocks by number of sides and corners.

Name_____ Date_____

2-DIMENSIONAL FIGURES
Sides and Corners

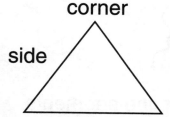

A triangle has 3 sides and 3 corners.

A rectangle has 4 sides and 4 square corners.

Write how many sides, corners, and square corners.

1.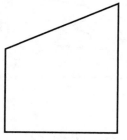

_4_ sides
_4_ corners
_2_ square corners

2.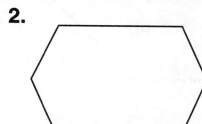

____ sides
____ corners
____ square corners

3.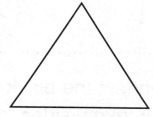

____ sides
____ corners
____ square corners

4.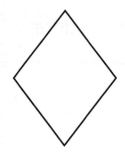

____ sides
____ corners
____ square corners

5.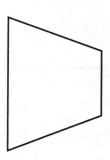

____ sides
____ corners
____ square corners

6.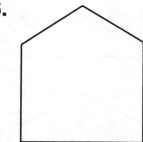

____ sides
____ corners
____ square corners

Match Attribute Blocks to shapes on page.

Name_____ Date _____

2-DIMENSIONAL FIGURES
Line Segments

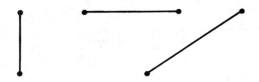   

These are line segments.   These are not line segments.

Is the figure a line segment?

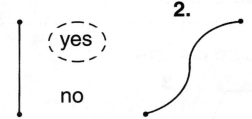

1. yes (circled)
   no

2. yes
   no

3. yes
   no

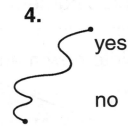

4. yes
   no

Connect the black dots.
How many sides, corners, and square corners did you draw?

5.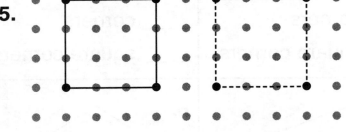

   __4__ sides
   __4__ corners
   __4__ square corners

6.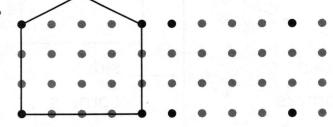

   ____ sides
   ____ corners
   ____ square corners

Duplicate shapes on Geoboards.

2-DIMENSIONAL FIGURES
Sides, Corners, and Line Segments

Is the figure a line segment? Ring yes or no.

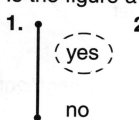 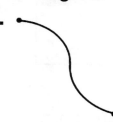 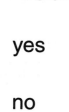 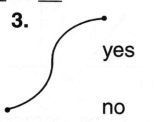 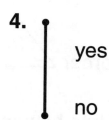

Connect the dots with line segments. How many sides, corners, and square corners are there?

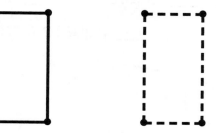

5. __4__ sides   __4__ corners
   __4__ square corners

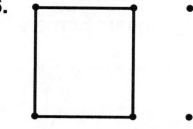

6. _____ sides   _____ corners
   _____ square corners

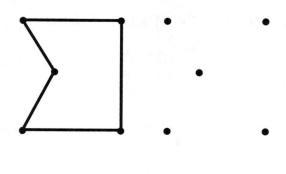

7. _____ sides   _____ corners
   _____ square corners

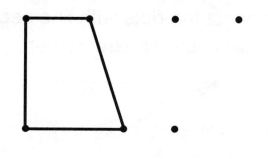

8. _____ sides   _____ corners
   _____ square corners

Sort Attribute Blocks by number of sides and corners.

Name _____ Date _____

2-DIMENSIONAL FIGURES
Sides, Corners, and Line Segments

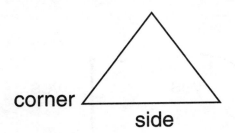
corner / side

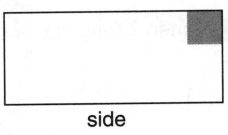
square corner / side

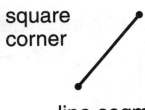
line segment

Write how many sides, corners, and square corners.

1.
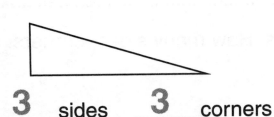

__3__ sides   __3__ corners
__1__ square corners

2.
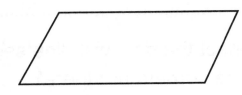

_____ sides   _____ corners
_____ square corners

3.
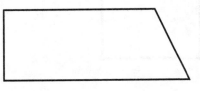

_____ sides   _____ corners
_____ square corners

4.
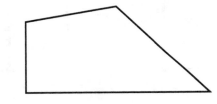

_____ sides   _____ corners
_____ square corners

Connect the dots with line segments. Write how many sides, corners, and square corners you drew.

5.

_____ sides   _____ corners
_____ square corners

6.
•
     •   •

_____ sides   _____ corners
_____ square corners

Sort Attribute Blocks by number of sides and corners. ⬜

2-DIMENSIONAL FIGURES
Sides and Corners

1. Draw a shape that has 8 sides and 8 corners.

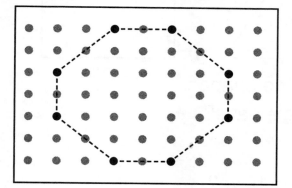

2. Draw a shape that has 5 sides and 5 corners.

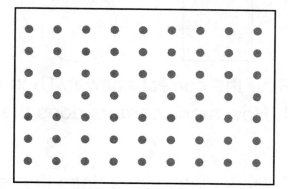

3. Draw a shape that has 3 sides and 3 corners.

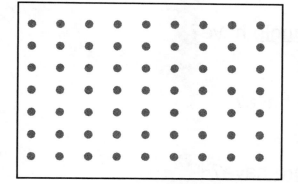

4. Draw a shape that has 4 sides and 4 corners. All 4 sides are the same length.

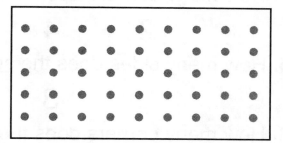

Visual Thinking

5. In this box, draw a shape that has sides and corners. Cover it. Describe it to a friend. Have your friend draw it. Compare your shape with the one your friend drew.

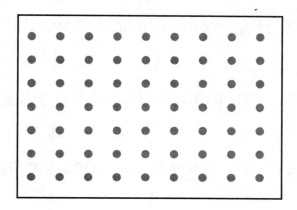

Duplicate shapes on Geoboards.

Name _____   Date _____

2-DIMENSIONAL FIGURES
Plane Figures

square   circle   triangle   rectangle

Ring the correct answer. Then color the shapes.

1. How many corners does the <u>triangle</u> have?

    2   ③   4

2. How many sides does the <u>rectangle</u> have?

    2   3   4

3. How many straight sides does the <u>circle</u> have?

    0   4   2

4. How many corners does the <u>rectangle</u> have?

    2   4   6

5. How many sides does the <u>square</u> have?

    2   3   4

6. How many corners does the <u>square</u> have?

    2   3   4

7. How many corners does the <u>circle</u> have?

    0   1   2

8. How many sides of the <u>square</u> are the same length?

    3   4   5

9. How many sides does the <u>triangle</u> have?

    1   2   3

Check answers with Overhead Attribute Blocks.

Name _____   Date _____

2-DIMENSIONAL FIGURES

Plane Figures

Ring each shape that has five sides. Then draw two more shapes that have five sides.

1.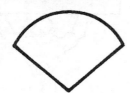

Ring each shape that has six sides. Then draw two more shapes that have six sides.

2.

Ring each shape that has equal sides. Then draw two more shapes that have equal sides.

3.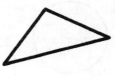

Ring each shape that has an odd number of sides. Then draw two more shapes that have an odd number of sides.

4.

Duplicate shapes on Geoboards.

Name_____ Date_____

2-DIMENSIONAL FIGURES
Plane Figures

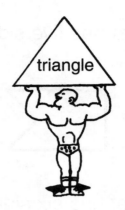
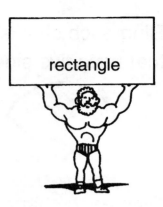

Write how many there are of each shape.

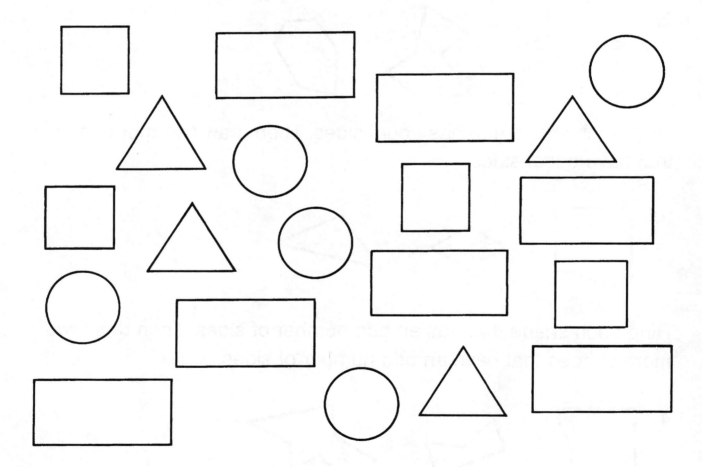

_____ circles _____ squares _____ triangles _____ rectangles

Sort Attribute Blocks by shape.

Name_____ Date _____

2-DIMENSIONAL FIGURES
Venn Diagram

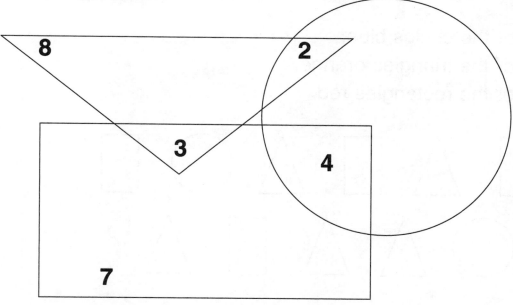

Use the shapes to answer each question. Write the answer.

1. Which number is in both the ○ and the ▽? _____

2. Which number is in both the ○ and the ▭? _____

3. Which number is in both the ▽ and the ▭? _____

4. What is the sum of the numbers in the ○ ? _____

5. What is the sum of the numbers in the ▽ ? _____

6. What is the sum of the numbers in the ▭ ? _____

7. Which shape has the highest sum? _____

8. Which shape has the lowest sum? _____

Name _____ Date _____

2-DIMENSIONAL FIGURES

Identifying Plane Figures

1. Color the circles blue.
   Color the triangles orange.
   Color the rectangles red.

2. Count the shapes. Write how many of each.

   __2__ circles   _____ triangles   _____ rectangles

3. Color the triangles purple.
   Color the circles red.
   Color the squares green.

4. Count the shapes. Write how many of each.
   _____ triangles   _____ circles   _____ squares

Sort and count Attribute Blocks by shape.

Name_____ Date_____

2-DIMENSIONAL FIGURES

Identifying Plane Figures

1. Color inside the triangles. Use blue.
   Color inside the circles. Use orange.
   Color inside the rectangles. Use green.

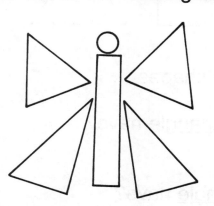 

Count each shape.
Write how many.

_____ triangles   _____ circles   _____ rectangles

....

2. Color inside the squares. Use blue.
   Color inside the triangles. Use orange.
   Color inside the circles. Use purple.

 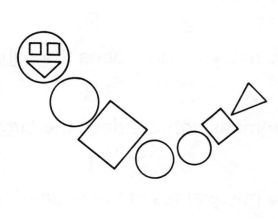

Count each shape. Write how many.

_____ squares   _____ triangles   _____ circles

2-DIMENSIONAL FIGURES

Attributes of Plane Figures

circle　　　square　　　rectangle　　　triangle

Ring the correct answer. Then color the shapes.

1. How many square corners does the <u>triangle</u> have?

   0　　　2　　　3

2. How many long sides does the <u>rectangle</u> have?

   2　　　3　　　4

3. How many straight sides does the <u>circle</u> have?

   0　　　2　　　4

4. How many square corners does the <u>rectangle</u> have?

   2　　　4　　　6

5. How many sides of the <u>triangle</u> are the same length?

   2　　　3　　　4

6. How many corners does the <u>square</u> have?

   2　　　3　　　4

7. How many corners does the <u>circle</u> have?

   0　　　1　　　2

8. How many sides of the <u>square</u> are the same length?

   3　　　4　　　5

Check answers with Overhead Attribute Blocks.

Name _____ Date _____

2-DIMENSIONAL FIGURES
Drawing Plane Figures

Look at each figure.
Draw a figure that is the same size and shape.

1.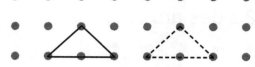

2.

3.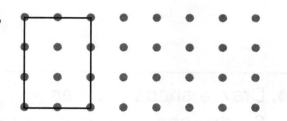

4.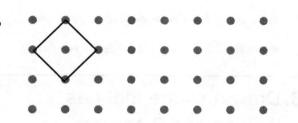

5.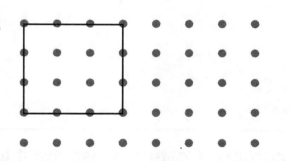

6.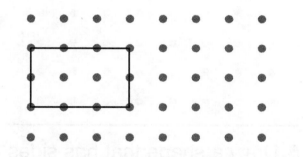

7. Draw a shape that has 3 sides and 3 corners.

Duplicate shapes on Geoboards.

www.svschoolsupply.com
© Steck-Vaughn Company

23

Unit 1: 2-Dimensional Figures
Geometry 2, SV 5806-X

Name _____ Date _____

2-DIMENSIONAL FIGURES
Drawing Plane Figures

1. Draw a shape that has 4 sides and 4 corners. All 4 sides are the same length.

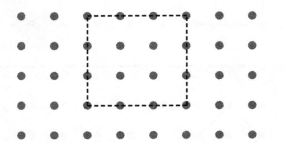

2. Draw a shape that has 4 sides and 4 corners. 2 sides are long. 2 sides are short.

3. Draw a shape that has 3 sides and 3 corners.

4. Draw a shape that has 6 sides and 6 corners.

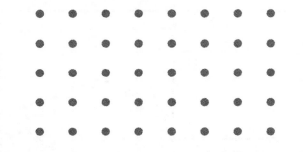

5. Draw a shape that has sides and corners. Cover it. Describe it to your friend. Have your friend draw it. Compare your shape with the one your friend drew.

your shape

your friend's shape

Duplicate shapes on Graph Paper.

www.svschoolsupply.com
© Steck-Vaughn Company

24

Unit 1: 2-Dimensional Figures
Geometry 2, SV 5806-X

Name _____ Date _____

2-DIMENSIONAL FIGURES
Drawing Plane Figures

1. Draw a shape that has 8 sides and 8 corners.

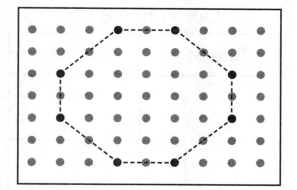

2. Draw a shape that has 3 sides and 3 corners.

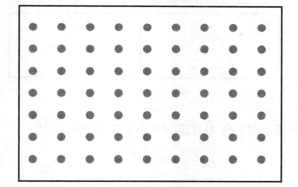

3. Draw a shape that has 4 sides and 4 corners. All 4 sides are the same length.

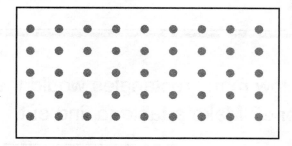

4. In this box, draw a shape that has sides and corners. Cover it. Describe it to a friend. Have your friend draw it. Compare your shape with the one your friend draws.

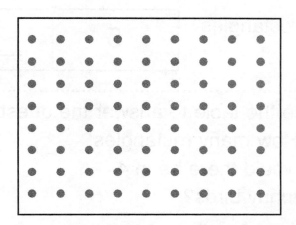

Match drawings to Attribute Blocks.

2-DIMENSIONAL FIGURES

Making a Table

1. Some children made this funny bird. How many of each shape did they use? Complete the table to find out.

Our Funny Bird

Shape	Number
○	
▭	
▭	
△	

2. How many rectangles would there be in 4 funny birds like this one? Make a table to find out.

birds	1	2	3	4
rectangles	5	___	___	___

Use the table to answer the questions.

3. How many rectangles would there be in 4 funny birds? ___

4. How many rectangles would there be in 3 funny birds? ___

Make funny animals with Tangrams.

Name _____ Date _____

2-DIMENSIONAL FIGURES
Problem Solving

Answer the questions.

  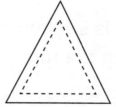 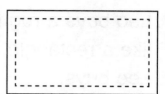

1. How many straight sides does the **circle** have? ____
   How many corners does it have? ____

2. How many sides does the **triangle** have? ____
   How many corners does it have? ____

3. How many sides does the **square** have? ____
   How many sides are the same length? ____
   How many corners does it have? ____

4. How many sides does the **rectangle** have? ____
   How many corners does it have? ____

5. How many triangles can you find?

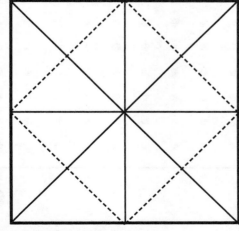

____ without the dotted lines
____ with the dotted lines

Check answers with Overhead Attribute Blocks.

www.svschoolsupply.com
© Steck-Vaughn Company

27

Unit 1: 2-Dimensional Figures
Geometry 2, SV 5806-X

Name _____ Date _____

2-DIMENSIONAL FIGURES

Problem Solving

Follow the instructions.

1. Leo buys a rug that is shaped like a rectangle. Ring the rug Leo buys.

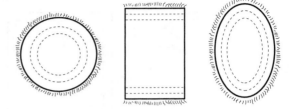

2. Anna uses shapes to make this figure. How many squares does she use?

_____

3. Albert folds a sheet of paper. He cuts out a shape. Ring the shape Albert cuts.

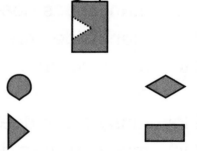

4. Pam drew a shape that has 5 sides and 5 corners. Draw the shape.

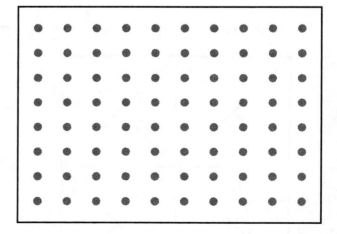

www.svschoolsupply.com
© Steck-Vaughn Company

28

Unit 1: 2-Dimensional Figures
Geometry 2, SV 5806-X

Name _____ Date _____

3-DIMENSIONAL FIGURES

Assessment: Unit 2

1. Count each shape. Write how many.

2. Ring the shape of the object's top.

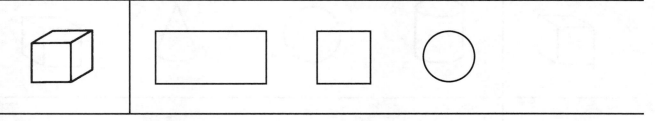

3. Ring the shape that has no top or bottom.

Name _____ Date _____

3-DIMENSIONAL FIGURES
Solids

sphere   cone   cylinder   cube   rectangular prism

Ring the same shape.

1.

2.

3.

4.

5.

Match shapes to Wooden or Plastic Geometric Solids.

3-DIMENSIONAL FIGURES
Solids

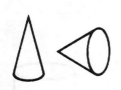 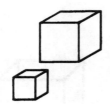 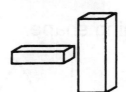 

cones   cubes   cylinders   rectangular prisms   spheres

Ring the matching shape.

1.

2.

3.

4. Match.

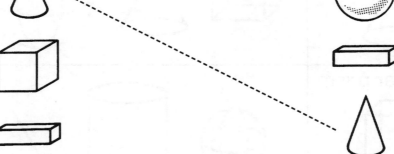
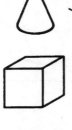

Match shapes to Wooden or Plastic Geometric Solids.

Name _____ Date _____

3-DIMENSIONAL FIGURES
Solids

Ring the matching shape.

1. sphere

  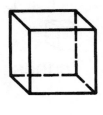 

2. cylinder

3. cube

    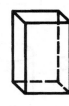

4. cone

   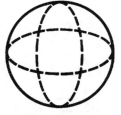 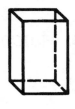

5. rectangular prism

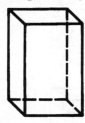   

Match shapes to Wooden or Plastic Geometric Solids.

3-DIMENSIONAL FIGURES
Solids and Plane Figures

Ring each matching object.

1.

2.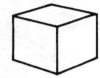

3.

4.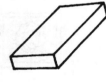

5.

6.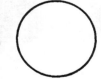

Match shapes to objects in classroom.

Name _____ Date _____

3-DIMENSIONAL FIGURES
Solids and Plane Figures

Ring each matching object.

1.	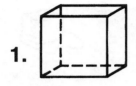				
2.					
3.					

Ring each matching shape.

4.					
5.					
6.					

Match shapes to objects in classroom.

Name_____ Date _____

3-DIMENSIONAL FIGURES

Identifying Solids

1. Color each ◯ red.   Color each ▭ purple.
   Color each ◻ green.   Color each ▭ brown.
   Color each △ blue.   Color each △ orange.

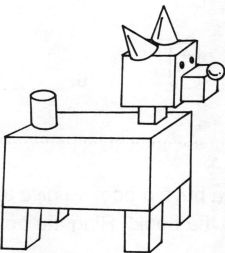

2. Count each shape. Write how many.

   __5__ ◯   ____ ▭
   ____ ◻   ____ ▭
   ____ △   ____ △

Sort and count Wooden or Plastic Geometric Solids.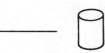

Name_____ Date_____

3-DIMENSIONAL FIGURES
Identifying Shapes

Color each shape.

1.

　　green　　　　yellow　　　　orange

　　purple　　　　red　　　　　blue

....

Study each shape. Write how many sides are showing.

2. _3_ 🔲   3. ___ ▯   4. ___ △

5. ___ ⌭   6. ___ ◯   7. ___ △

....

8. Maria buys a book to give to her friend. In which box did she wrap the book? Ring the box.

Compare number of sides of Attribute Blocks with Geometric Solids.

Name _____ Date _____

3-DIMENSIONAL FIGURES
Attributes of Solids

Each of the shapes has an underside.

 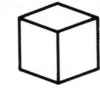

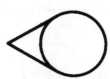 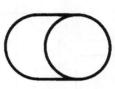

Draw the underside for each shape.

1.   2.

3.   4.

A cone can stack on a cube.

Think about each shape. Write the answer.

5. Can a stack on a ? _____

6. Can a stack on a ? _____

7. Can a stack on a ? _____

8. Can a stack on a ? _____

Check answers with Wooden or Plastic Geometric Solids.

Name _____ Date _____

3-DIMENSIONAL FIGURES
Attributes of Solids

Ring the shape of each object's top.

1.

2.

3.

4.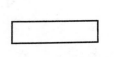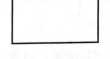

5.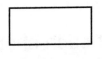

...g the shape that has no
...or bottom.

Check answers with Wooden or Plastic Geometric Solids.

Unit 2: 3-Dimensional Figures
Geometry 2, SV 5806-X

Name _____ Date _____

3-DIMENSIONAL FIGURES
Problem Solving

Jessica, Luis, and Alexis are learning about solids. Help them answer the questions.

1. Can the **sphere** roll? __yes__
   How many flat sides does it have? __0__
   How many corners does it have? __0__

2. Can the **cone** roll? _____
   Can it slide? _____
   How many flat sides does it have? _____
   How many corners does it have? _____

3. Can the **cylinder** roll? _____
   How many flat sides does it have? _____
   How many corners does it have? _____

4. Can the **cube** roll? _____
   Can it slide? _____
   How many flat sides does it have? _____
   How many corners does it have? _____

5. Can the **rectangular prism** roll? _____
   Can it slide? _____
   How many flat sides does it have? _____
   How many corners does it have? _____

Answer questions using Wooden or Plastic Geometric Solids.

CONGRUENCE
Assessment: Unit 3

Ring the one that is the same shape and size.

1.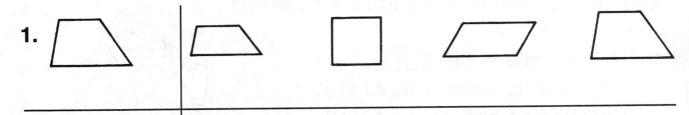

2.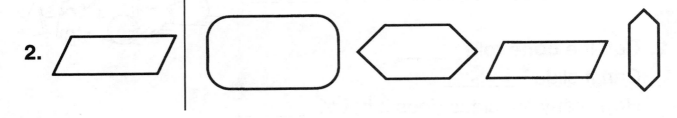

Draw a figure that is the same size and shape.

3.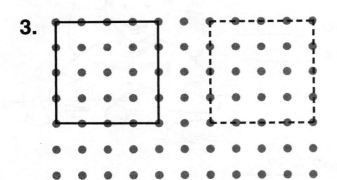

4.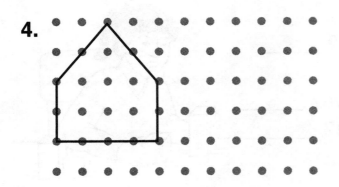

Name_____ Date_____

CONGRUENCE

Matching Shapes

Ring the one that is the same shape and size.

1. |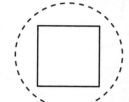

2. |

3.  |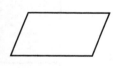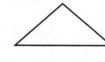

4. 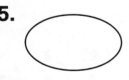 |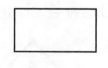

5.  |

6.

Find congruent shapes using Pattern Blocks.

Name _____   Date _____

CONGRUENCE
Same Shape, Same Size

same shape
same size

same shape
not the same size

....

Ring the one that is the same shape and size.

1. |

2.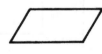

3. |

....

Draw a shape that is the same size.

4. 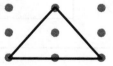    5.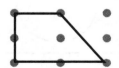

6.    7.

Make congruent shapes on Geoboards.

Name_____ Date _____

CONGRUENCE
Same Shape, Same Size

Ring the one that is the same shape and size.

1.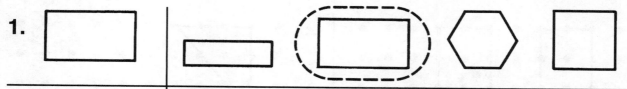

2.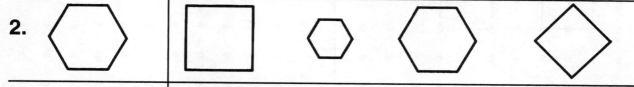

3.

....

Draw a shape that is the same size.

4.    5.

6. 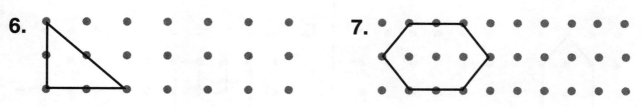   7.

8.    9.

Make congruent shapes on Graph Paper.

Name _____   Date _____

CONGRUENCE

Drawing Congruent Shapes

Draw a shape that is the same size.

1.

2.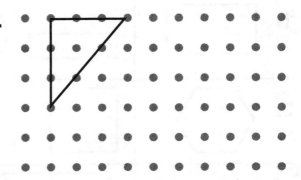

3.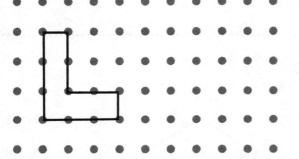

4.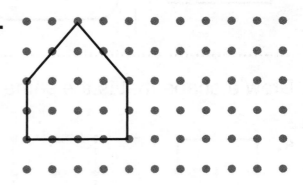

5.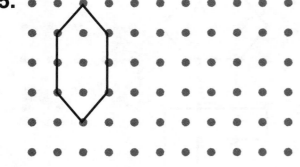

6.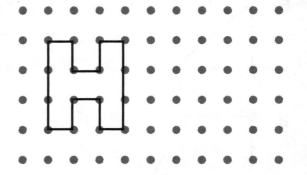

Make congruent shapes on Geoboards.

www.svschoolsupply.com
© Steck-Vaughn Company

44

Unit 3: Congruence
Geometry 2, SV 5806-X

CONGRUENCE
Drawing Congruent Figures

Draw a shape that is the same size.

1.

2.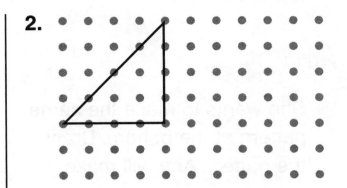

3.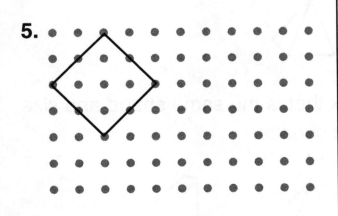

4.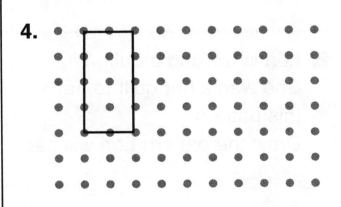

5.

6.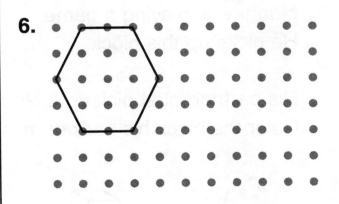

Make congruent shapes on Geoboards.

CONGRUENCE
Problem Solving

1. Ana makes this pattern at the top of her paper.

   She wants to make the same pattern at the bottom. Draw the pattern Ana will make.

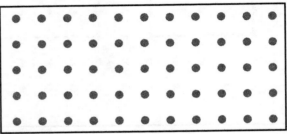

2. Lori is making a quilt. She wants her quilt to have this pattern.
   Draw the pattern Lori will use.

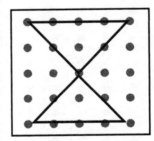 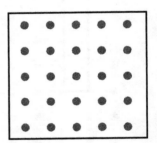

3. Matthew is playing a game. He picks up this block.

   His partner must pick up a block that is the same shape and size. Color the block his partner must choose.

Play Matthew's game using two sets of Geometric Solids.

Name _____ Date _____

SYMMETRY

Assessment: Unit 4

Is the line a line of symmetry? Ring <u>yes</u> or <u>no</u>.

1. 　　2. 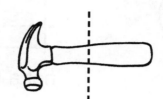　　3.

　　yes　(no)　　　　yes　no　　　　yes　no

Draw one line of symmetry for each shape.

4. 　　5. 　　6.

Draw two lines of symmetry for each shape.

7. 　　8. 　　9.

www.svschoolsupply.com　　　　47　　　　Unit 4: Symmetry
© Steck-Vaughn Company　　　　　　　　Geometry 2, SV 5806-X

Name _____ Date _____

SYMMETRY
Identifying Symmetry

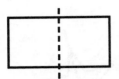 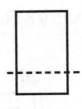 

line of symmetry   not a line of symmetry

Is the line a line of symmetry? Ring yes or no.

1. 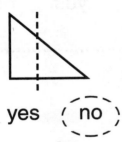
yes   (no)

2. 
yes   no

3. 
yes   no

4. 
yes   no

5. 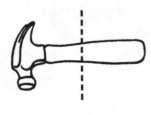
yes   no

6. 
yes   no

Draw a line of symmetry.

7.

8.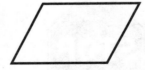

9.

10.

11.

Trace shape and fold paper to find line of symmetry. ◻

www.svschoolsupply.com
© Steck-Vaughn Company

48

Unit 4: Symmetry
Geometry 2, SV 5806-X

Name _____ Date _____

SYMMETRY
Identifying Symmetry

Is the line a line of symmetry? Ring <u>yes</u> or <u>no</u>.

1.

   yes   no

2.

   yes   no

3.

   yes   no

4.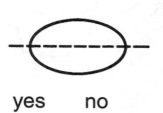

   yes   no

5.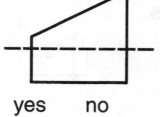

   yes   no

6.

   yes   no

7.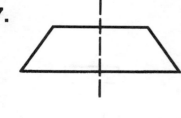

   yes   no

8.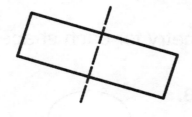

   yes   no

9.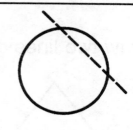

   yes   no

Draw a line of symmetry for each shape.

10.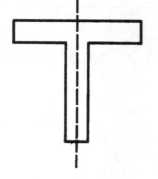

11.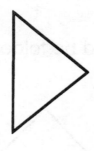

12.

Trace shape and fold paper to find line of symmetry.

Name _____ Date _____

SYMMETRY
Drawing Lines of Symmetry

Draw one line of symmetry for each shape.

1.    2.   3.

4.    5.   6.

Draw two lines of symmetry for each shape.

7. 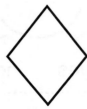   8. 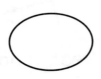   9.

10. Ring the shape if it could be folded to make a line of symmetry.

Trace shape and fold paper to find line of symmetry. ⬜

www.svschoolsupply.com
© Steck-Vaughn Company

50

Unit 4: Symmetry
Geometry 2, SV 5806-X

Name_____ Date _____

SYMMETRY
Drawing Lines of Symmetry

Draw one line of symmetry for each shape.

1.
2.
3.

4.
5.

Draw two lines of symmetry for each shape.

6.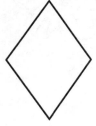
7.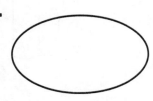
8.

9.
10.
11.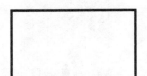

Trace shape and fold paper to find line of symmetry.

SYMMETRY
Problem Solving

1. Mark has these sheets of paper. Which can he fold so the new shapes look the same? Circle the shapes.

···

2. Mark doesn't think he can fold these shapes to make lines of symmetry. Help him by drawing one line of symmetry for each shape.

   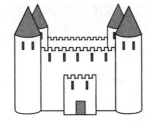   

PERIMETER AND AREA
Assessment: Unit 5

Add the lengths. Write the perimeter.

1. 2 cm
    1 cm [ ] 1 cm   ___ + ___ + ___ + ___ = ___ cm
    2 cm

2. 3 cm
    1 cm [ ] 1 cm   ___ + ___ + ___ + ___ = ___ cm
    3 cm

Use your inch ruler. Measure the sides of each figure. Find the perimeter.

3. 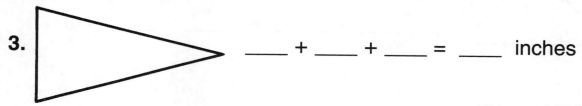   ___ + ___ + ___ = ___ inches

4. 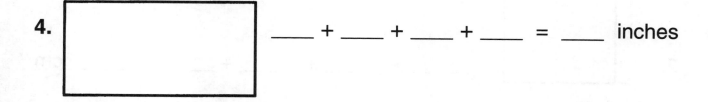   ___ + ___ + ___ + ___ = ___ inches

Count the square units. Write the number of square centimeters.

5.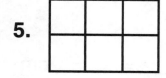

6. [grid]   ___ square centimeters

square centimeters ___

Name _____ Date _____

PERIMETER AND AREA
Measuring Perimeter

To find the perimeter of a shape, measure each side. Then add.

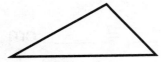

3 cm + 2 cm + 4 cm = 9 cm
The perimeter is 9 cm.

....

Use your centimeter ruler.
Measure each side.
Add the lengths.

1. 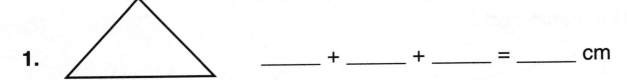   ____ + ____ + ____ = ____ cm

2.    ____ + ____ + ____ + ____ = ____ cm

3. 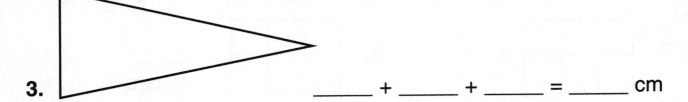   ____ + ____ + ____ = ____ cm

Measure sides with Centimeter Rulers.

www.svschoolsupply.com
© Steck-Vaughn Company

Unit 5: Perimeter and Area
Geometry 2, SV 5806-X

PERIMETER AND AREA

Measuring Perimeter

Measure each side of the triangle with a centimeter ruler. Then add to find the perimeter.

1.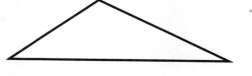

   __4__ + __3__ + __6__ = __13__ centimeters

   The perimeter of the triangle is __13__ centimeters.

2.

   _____ + _____ + _____ + _____ = _____ centimeters

3.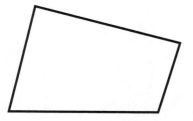

   _____ + _____ + _____ + _____ = _____ centimeters

   Measure sides with Centimeter Rulers.

Name_____ Date _____

PERIMETER AND AREA
Measuring Perimeter

Use your centimeter ruler to measure the perimeter.

1.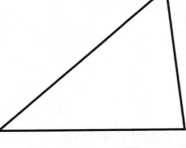

___6___ + ___5___ + ___4___ = ___15___ centimeters

The perimeter of the triangle is ___15___ centimeters.

2.

_____ + _____ + _____ + _____ = _____ centimeters

3.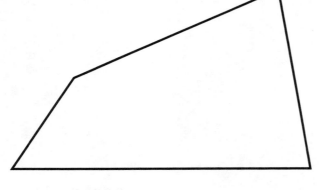

_____ + _____ + _____ + _____ = _____ centimeters

Measure sides with Centimeter Rulers.

www.svschoolsupply.com
© Steck-Vaughn Company

Unit 5: Perimeter and Area
Geometry 2, SV 5806-X

Name_____ Date_____

PERIMETER AND AREA
Measuring Perimeter

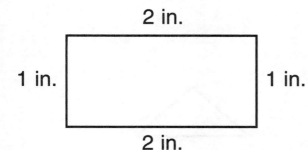

1 + 2 + 1 + 2 = 6 in.

Add the lengths of the sides to find the perimeter.

Measure the sides of each figure.
Find the perimeter. Write the number of inches.

....

1. 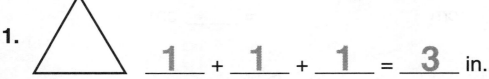   __1__ + __1__ + __1__ = __3__ in.

2.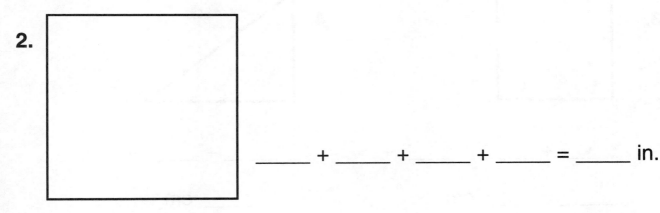

   ____ + ____ + ____ + ____ = ____ in.

3.

   ____ + ____ + ____ + ____ = ____ in.

4.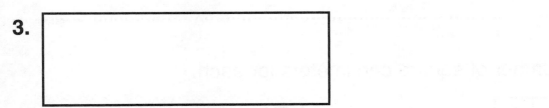

____ + ____ + ____ = ____ in.   Measure sides with Inch Rulers.

www.svschoolsupply.com
© Steck-Vaughn Company

57

Unit 5: Perimeter and Area
Geometry 2, SV 5806-X

Name _____ Date _____

PERIMETER AND AREA

Perimeter/Area

Use your centimeter ruler.
Measure each side. Add the lengths.

1.

2.

___ + ___ + ___ = ___ cm

3.

___ + ___ + ___ + ___ = ___ cm

4.

___ + ___ + ___ = ___ cm

Write the number of square centimeters for each.

5.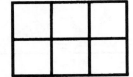

_6_ square centimeters

6.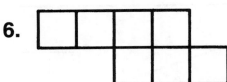

____ square centimeters

Measure sides with Centimeter Rulers.

Name _____ Date _____

PERIMETER AND AREA
Perimeter/Area

Use your inch ruler. Measure the sides of each figure. Find the perimeter.

1.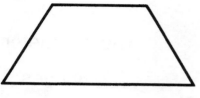

____2____ + _____ + _____ + _____ = _____ inches

2.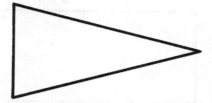

_____ + _____ + _____ = _____ inches

3.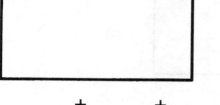

_____ + _____ + _____ + _____ = _____ inches

4.

_____ + _____ + _____ + _____ = _____ inches

....

Write the number of square inches.

5.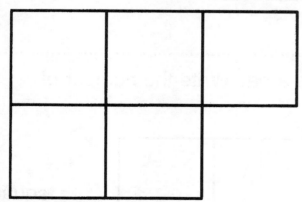

____5____ square inches

6.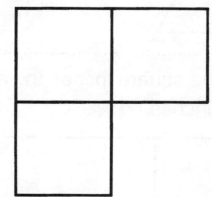

_____ square inches

Measure sides with Inch Rulers.

Name _____ Date _____

PERIMETER AND AREA

Perimeter/Area

Add the lengths of the sides to find the perimeter.

1. 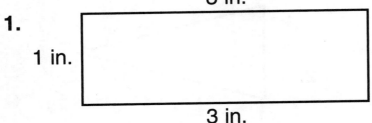   The perimeter is ____ inches.

   ____ + ____ + ____ + ____ = ____ inches

2. 4 in.

   1 in.   1 in.

   4 in.

   ____ + ____ + ____ + ____ = ____ inches

3. △   ____ + ____ + ____ = ____ inches

Count the square inches to find the area. Write the number of square inches.

4. 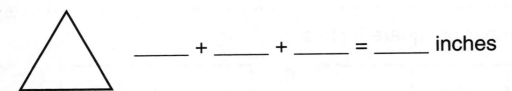   ____ square inches

Measure sides with Inch Rulers.

PERIMETER AND AREA
Problem Solving

There is a fence around Spot's backyard. The tree is 5 meters away from Spot's water dish. The tree is 7 meters away from Spot's doghouse.

1. Write each fence length.

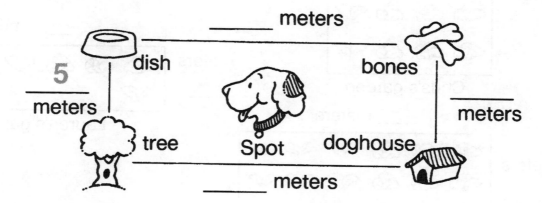

2. Spot buried some bones. Then he drank some water from his water dish. How far did he walk from the bones to the water dish? _____ meters

3. Later, Spot walked from his doghouse back to where the bones were buried. How far did he walk? _____ meters

4. What numbers would you add to show how far it is around Spot's yard?

   _7_ + ____ + ____ + ____

5. How many meters is it around Spot's yard? _____ meters

PERIMETER AND AREA
Problem Solving

Every 🥬 needs land that is 1 meter on each side.

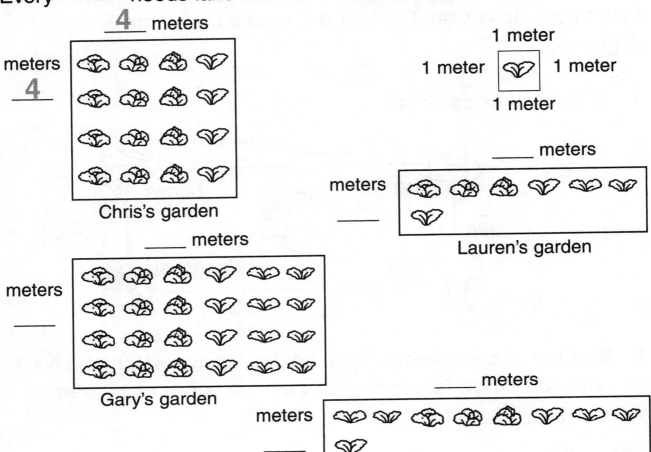

1. Write the meters on each side of the gardens.

2. Draw the rest of the 🥬 in each garden.

Add the sides of each garden. Write the total.

3. Chris' garden ___16___ meters   4. Lauren's garden _____ meters

5. Gary's garden _____ meters   6. Jo's garden _____ meters

7. Whose garden has the most ? _____

PERIMETER AND AREA
Perimeter Using Scale

In this rectangle, each ☐ is 50 feet on each side.

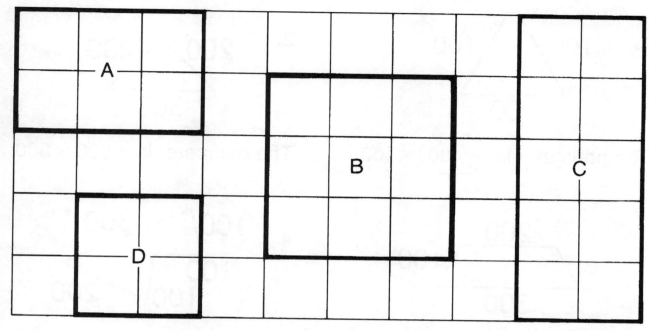

Look at each shape. Count and add to write the answer.

1. What is the distance around A?
   _____

2. What is the distance around B?
   _____

3. What is the distance around C?
   _____

4. What is the distance around D?
   _____

www.svschoolsupply.com
© Steck-Vaughn Company

63

Unit 5: Perimeter and Area
Geometry 2, SV 5806-X

Name _____ Date _____

PERIMETER AND AREA
Estimating Perimeter

Look at each shape. Ring your estimate.

1.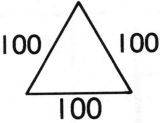

The perimeter is (< 400) > 400.

2.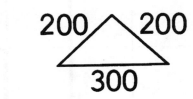

The perimeter is < 500 > 500.

3.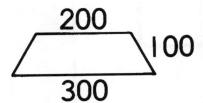

The perimeter is < 500 > 500.

4.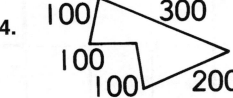

The perimeter is < 900 > 900.

These shapes have equal sides. Ring your guess.

5.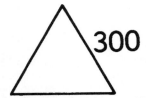

The perimeter is < 800 > 800.

6.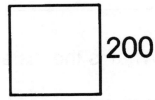

The perimeter is < 400 > 400.

7.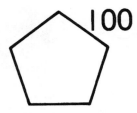

The perimeter is < 400 > 400.

8.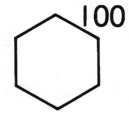

The perimeter is < 800 > 800.

PERIMETER AND AREA
Using a Calculator to Find Perimeter

Use a centimeter ruler to measure each shape.

Use a to find the perimeter. Write the perimeter.

1.

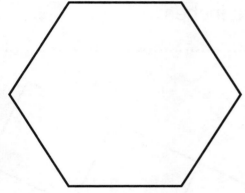

_____ centimeters

2.

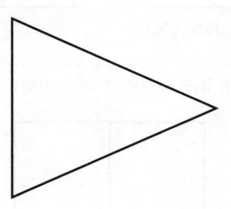

_____ centimeters

3.

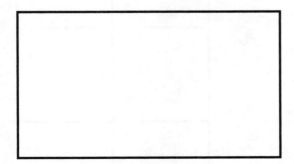

_____ centimeters

4.

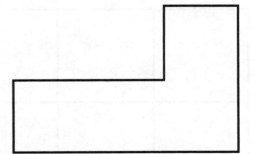

_____ centimeters

Find perimeter using a Calculator.

Name _____ Date _____

PERIMETER AND AREA

Area

Count square inches to find the area of a shape.

1 square inch   4 square inches

....

Write the number of square inches.

1.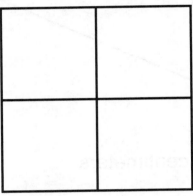

   _4_ square inches

2.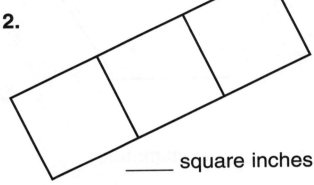

   ____ square inches

3.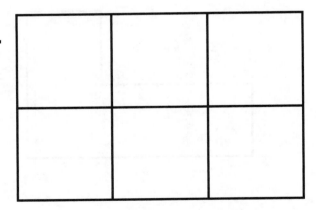

   ____ square inches

4.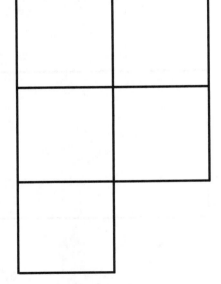

   ____ square inches

Check answers with 1-inch Counting Cubes.

www.svschoolsupply.com
© Steck-Vaughn Company

66

Unit 5: Perimeter and Area
Geometry 2, SV 5806-X

PERIMETER AND AREA

Area

To find the area of a shape, count the square units.

1 square centimeter

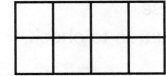

8 square centimeters

....

Write the number of square centimeters.

1.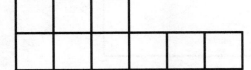

__9__ square centimeters

2.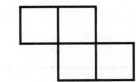

_____ square centimeters

3.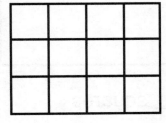

_____ square centimeters

4.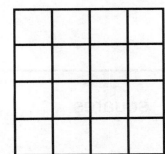

_____ square centimeters

5.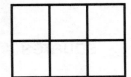

_____ square centimeters

6.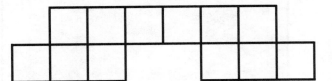

_____ square centimeters

Make other problems on Centimeter Graph Paper.

PERIMETER AND AREA

Problem Solving

How many 1-inch squares will fit in each space?
First, write your guess.
Then use 1-inch squares to check.

1. [rectangle]

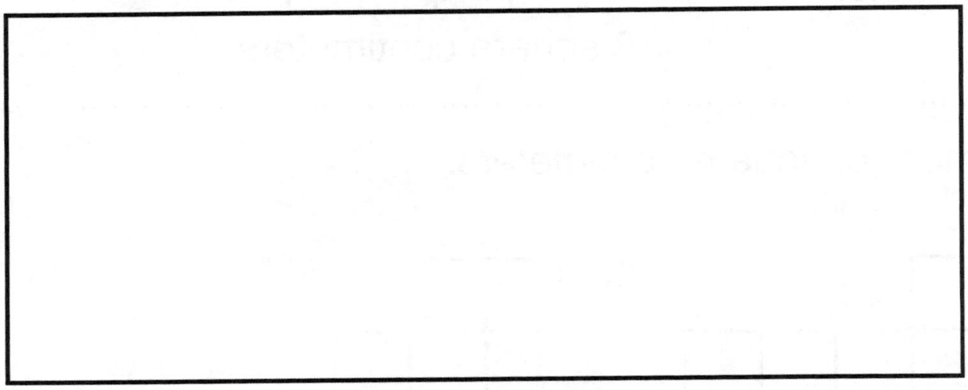

   Guess. _____ squares   Check. __10__ squares

2. [rectangle]

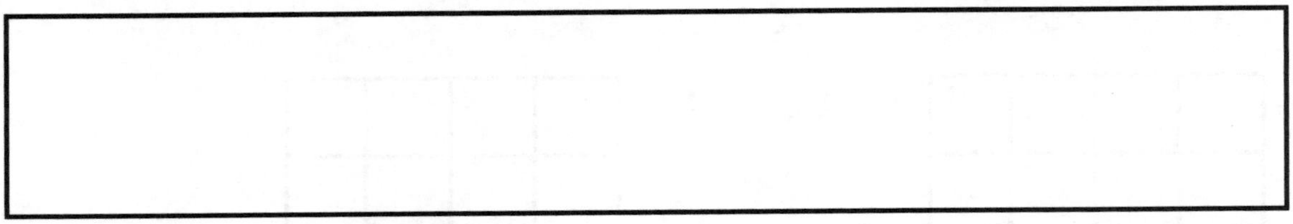

   Guess. _____ squares   Check. _____ squares

3. [rectangle]

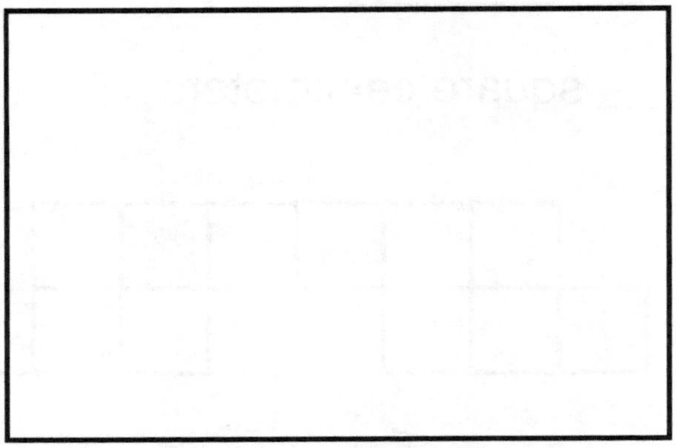

   Guess. _____ squares
   Check. _____ squares

Check answers with 1-inch Counting Cubes.

Name _____ Date _____

FRACTIONS USING PICTORIAL MODELS

Assessment: Unit 6

Write the number of equal parts.

1.
   _____

2.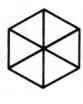
   _____

3.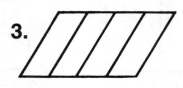
   _____

Write the fraction to show what part is shaded.

4.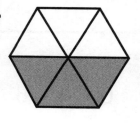
   _____

5.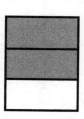
   _____

6.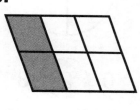
   _____

Answer <u>Yes</u> or <u>No</u>.

7. Rachel ate $\frac{1}{2}$ of a pear.
   Joseph ate $\frac{1}{4}$ of a pear.
   Did they eat the same amount? _____

FRACTIONS USING PICTORIAL MODELS

Meaning of Fractions

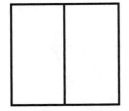
2 equal parts

3 equal parts

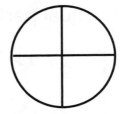
4 equal parts

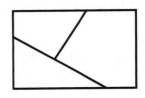
0 equal parts

Ring the number of equal parts.

1.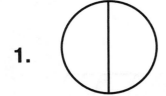
0 (2) 3

2.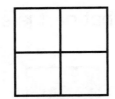
0   3   4

3.
0   4   6

4.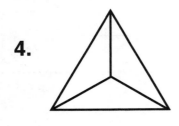
0   2   3

5.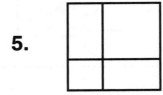
0   2   4

6.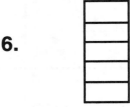
0   3   5

7.
0   4   8

8.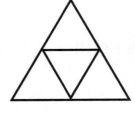
0   3   4

9.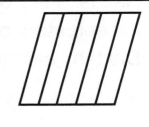
0   4   5

Demonstrate equal parts with Overhead Fraction Circles, Squares, and Rectangles.

Name _____ Date _____

FRACTIONS USING PICTORIAL MODELS
Meaning of Fractions

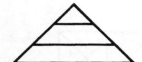 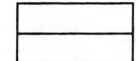 

0 equal parts   2 equal parts   8 equal parts   10 equal parts

Ring the number of equal parts.

1.
0 2 (3)

2.
0 3 6

3.
0 4 5

4.
0 1 2

5.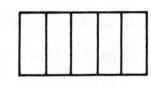
0 5 6

6.
0 4 8

Draw lines to show equal parts.

7. 6 equal parts   8. 4 equal parts   9. 3 equal parts

    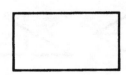   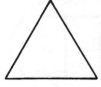

10. 2 equal parts   11. 5 equal parts   12. 8 equal parts

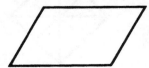

Duplicate equal parts with Overhead Fraction Circles, Squares, and Rectangles.

Name _____ Date _____

FRACTIONS USING PICTORIAL MODELS
Meaning of Fractions

Ring the number of equal parts.

1. 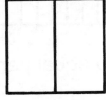   3   4   ⓶

2. 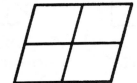   2   4   6

3. 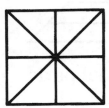   6   8   4

4. 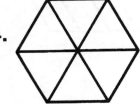   4   6   8

5. 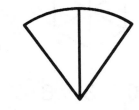   2   3   4

6.    6   7   8

Color each shape that shows equal parts.

7. 6 equal parts 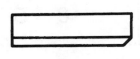

8. 4 equal parts 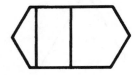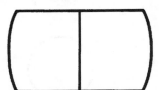

9. 8 equal parts 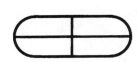

Duplicate equal parts on Jumbo Graph Paper.

FRACTIONS USING PICTORIAL MODELS

Identify Equal Parts

Color each shape that shows equal parts.

1. 3 equal parts

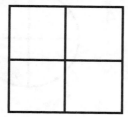 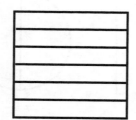 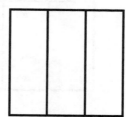 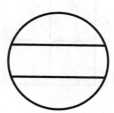

2. 8 equal parts

 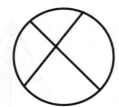 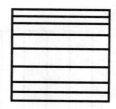 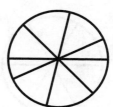

3. 10 equal parts

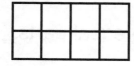 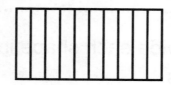 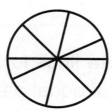 

Draw lines to show equal parts.

4. 2 equal parts   **5.** 4 equal parts   **6.** 6 equal parts

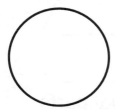 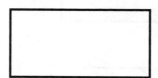 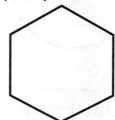

Demonstrate equal parts with Overhead Fraction Circles, Squares, and Rectangles.

Name_____ Date_____

FRACTIONS USING PICTORIAL MODELS

Equal Parts

Write the number of equal parts.

1. [square divided into 4]   2. 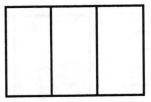   3. 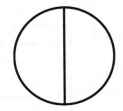   4.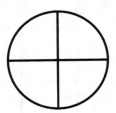

_____   _____   _____   _____

5. 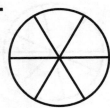   6. 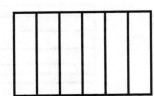   7. 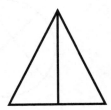   8.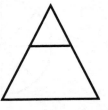

_____   _____   _____   _____

....

Use your ruler. Draw lines to divide each shape into equal parts. Write the number of equal parts.

9. 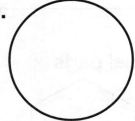   10.    11.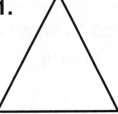

_____ equal parts   _____ equal parts   _____ equal parts

Count equal parts with Overhead Fraction Circles, Squares, and Rectangles.

FRACTIONS USING PICTORIAL MODELS

Equal Parts

Write the number of equal parts.

1.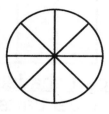
   _8_

2.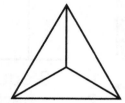
   ___

3.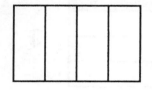
   ___

4.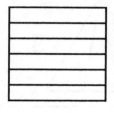
   ___

5.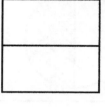
   ___

6.
   ___

7.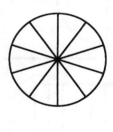
   ___

8.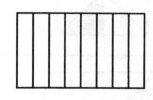
   ___

9. Four friends want to share two apples in equal parts. Show how they can divide them.

Demonstrate equal parts with Overhead Fraction Circles, Squares, and Rectangles.

Name_____ Date_____

FRACTIONS USING PICTORIAL MODELS

Fractions

halves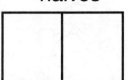
2 equal parts

halves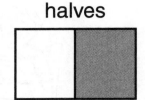
1 part shaded / 2 equal parts → 1/2

thirds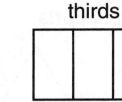
3 equal parts

thirds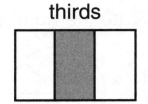
1 part shaded / 3 equal parts → 1/3

Complete each fraction to show what part is shaded.

1. 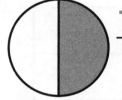   1 part shaded / 2 equal parts

2. 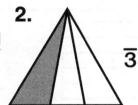 $\frac{}{3}$

3.  $\frac{}{4}$

4. 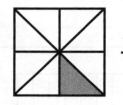 $\frac{1}{}$

5. 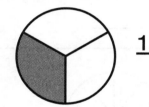 $\frac{1}{}$

6. 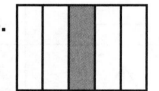 $\frac{1}{}$

7. 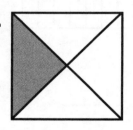 $\frac{}{4}$

8. 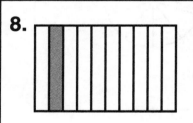 $\frac{1}{}$

9. 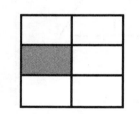 $\frac{1}{}$

Read each fraction. Shade the correct number of parts.

10.  $\frac{1}{3}$

11. 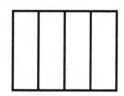 $\frac{1}{4}$

12.  $\frac{1}{8}$

Duplicate fractions with Fraction Builder Strips.

FRACTIONS USING PICTORIAL MODELS
Fractions

3 parts shaded → 3
5 equal parts   5
in all

Ring each fraction that tells what part is shaded.

1.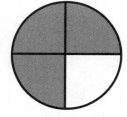

$\frac{1}{4}$ $\frac{2}{4}$ ⊙$\frac{3}{4}$⊙

2.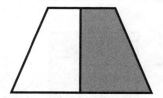

$\frac{1}{3}$ $\frac{1}{4}$ $\frac{1}{2}$

3.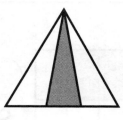

$\frac{1}{2}$ $\frac{1}{3}$ $\frac{1}{4}$

4.

$\frac{1}{2}$ $\frac{2}{3}$ $\frac{3}{5}$

5.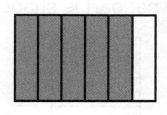

$\frac{1}{6}$ $\frac{4}{6}$ $\frac{5}{6}$

6.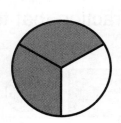

$\frac{1}{3}$ $\frac{2}{3}$ $\frac{3}{4}$

Complete each fraction to tell what part is shaded.

7.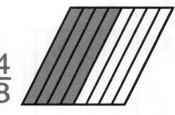

$\frac{4}{8}$

8.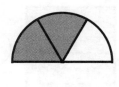

$\frac{}{3}$

9.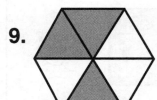

$\frac{3}{}$

Demonstrate fractions with Overhead Fraction Circles.

Name _____ Date _____

FRACTIONS USING PICTORIAL MODELS

Writing Fractions

Complete each fraction to tell what part is shaded.

1. 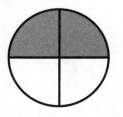 ②/4

2. 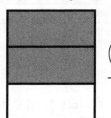 ○/3

3. 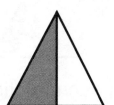 ○/2

4. 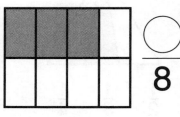 ○/8

5. 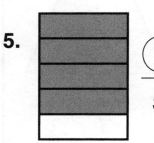 ○/5

6. 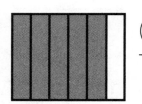 ○/6

Write a fraction that tells what part is shaded.

7. 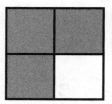 ③/④

8. 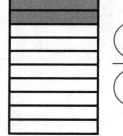 ○/○

9. 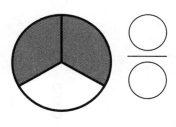 ○/○

10. 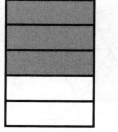 ○/○

11. 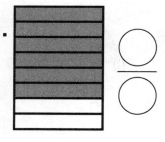 ○/○

12. 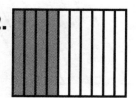 ○/○

Make and label fractions with Fraction Builder Strips.

Writing Fractions

FRACTIONS USING PICTORIAL MODELS

Complete each fraction so that it shows what part is shaded.

1. 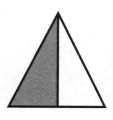 $\dfrac{1}{2}$

2. 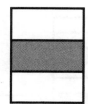 $\dfrac{}{3}$

3. 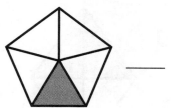 $\dfrac{}{}$

4. 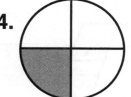 $\dfrac{1}{}$

5.  $\dfrac{}{}$

6.  $\dfrac{1}{}$

7.  $\dfrac{}{6}$

8. 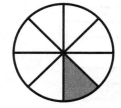 $\dfrac{1}{}$

9. 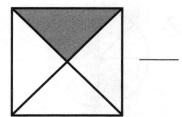 $\dfrac{}{}$

10. 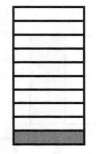 $\dfrac{}{}$

11. 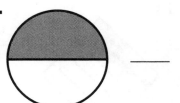 $\dfrac{}{}$

12. 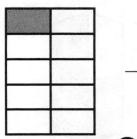 $\dfrac{1}{}$

Check answers using Overhead Fractions.

Name_____ Date_____

FRACTIONS USING PICTORIAL MODELS

Writing Fractions

Write the fraction to show what part is shaded.

1. 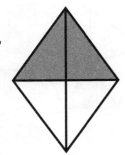 2

2.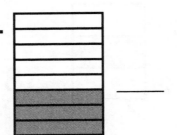

3.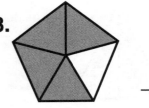

4.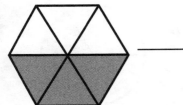

5.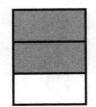

6.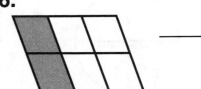

7.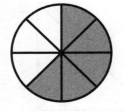

8.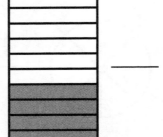

9.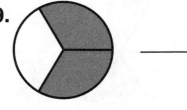

10.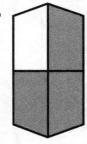

11.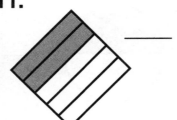

12.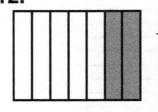

Duplicate fractions using Fraction Tiles.

Name_____ Date_____

FRACTIONS USING PICTORIAL MODELS
Writing Fractions

Complete a fraction to show what part is shaded.

1.

$$\frac{2}{6}$$

2.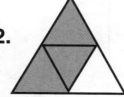

$$\frac{}{4}$$

3.

$$\frac{}{8}$$

4.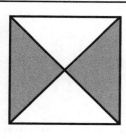

$$\frac{}{4}$$

5.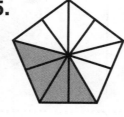

$$\frac{}{10}$$

6.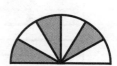

$$\frac{}{6}$$

Write a fraction to show what part is shaded.

7.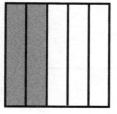

———

8.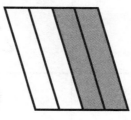

———

9.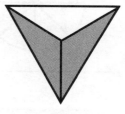

———

10.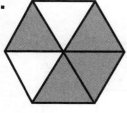

———

11.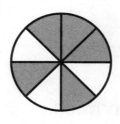

———

12.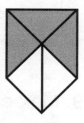

———

Duplicate fractions using Fractions Tiles.

www.svschoolsupply.com
© Steck-Vaughn Company

81

Unit 6: Fractions Using Pictorial Models
Geometry 2, SV 5806-X

Name_____ Date_____

FRACTIONS USING PICTORIAL MODELS

Writing Fractions

Complete each fraction to show what part is shaded.

1.  $\dfrac{1}{2}$

2. 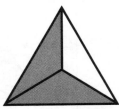 $\dfrac{}{3}$

3. 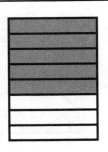 $\dfrac{}{8}$

4. 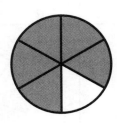 $\dfrac{}{6}$

Write a fraction to show what part is shaded.

5. 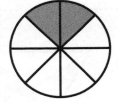 $\dfrac{2}{8}$

6. 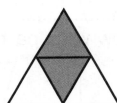 ___

7. 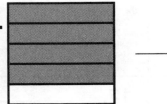 ___

8. 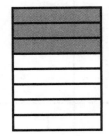 ___

9. Max eats $\dfrac{3}{4}$ of his sandwich.
Color the part of the sandwich Max eats.

Demonstrate fractions using Overhead Fraction Circles.

Name _____ Date _____

FRACTIONS USING PICTORIAL MODELS

Identifying Fractional Parts

Follow the instructions.

1. Color $\frac{1}{2}$.

 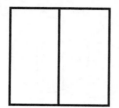 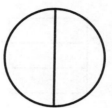 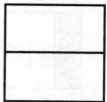

2. Color $\frac{1}{3}$.

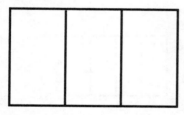 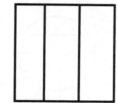  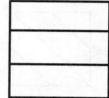

3. Color $\frac{1}{4}$.

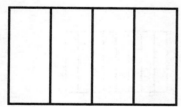 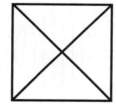 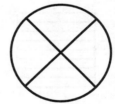 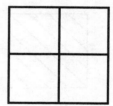

....

4. Solve.

Tony ate $\frac{1}{4}$ of a large pizza.

Rita ate $\frac{1}{4}$ of a medium pizza.

Did they eat the same amount? _____

Duplicate problem using Pizza Cardboards.

Name _____ Date _____

FRACTIONS USING PICTORIAL MODELS

Identifying Fractional Parts

Follow the instructions.

1. Color $\frac{1}{3}$.

2. Color $\frac{1}{4}$.

3. Color $\frac{1}{8}$.

---

4. The teacher asked Fran to fold her paper in sixths.
   Did Fran follow directions?
   Ring <u>Yes</u> or <u>No</u>.

   Yes   No

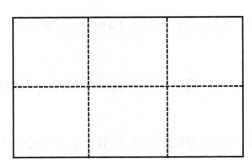

Duplicate Fran's problem by folding paper.

FRACTIONS USING PICTORIAL MODELS

Identifying Fractional Parts

Follow the instructions.

1. Color $\frac{1}{3}$.

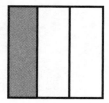 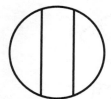  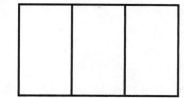

2. Color $\frac{1}{4}$.

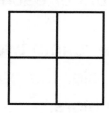 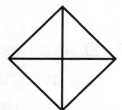 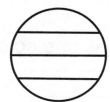 

3. Color $\frac{1}{8}$.

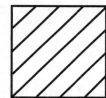 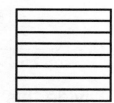 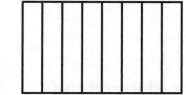 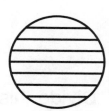

....

Ring **Yes** or **No**.

4. Tomiko colored $\frac{1}{2}$ of a circle.

   Selena colored $\frac{1}{2}$ of a larger circle.

   Did they color the same amount?

   Yes   No

Check answers using Overhead Fractions.

Name_____ Date _____

FRACTIONS USING PICTORIAL MODELS

Problem Solving

Carl is learning to build. He must know about fractions.

Answer the questions. Then write $\frac{1}{2}, \frac{1}{4}, \frac{3}{4}, \frac{1}{3}, \frac{2}{3}$, or 1 under each part.

1. Color this board all one color.

   _____
   __1__

2. Divide this board into two equal parts. Color the parts two different colors.

   _____
   ____ ____

3. Divide this board into three equal parts. Color one part with one color. Color the other two parts another color.

   _____
   ____ ____

4. Divide this board into four equal parts. Color one part. Color the other three parts with another color.

   _____
   ____ ____

5. Write the fractions from the least to the greatest. Write the whole number last.

   _____

PATTERNS AND COORDINATE GRAPHS

Assessment: Unit 7

Draw the shape that continues the pattern.

1.

2.

3. Ring the shape that continues the pattern.

   |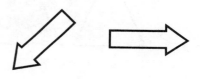

4. Start at 0. Go → 1. Go ↑ 3.
   Draw a ring at the crossing point.

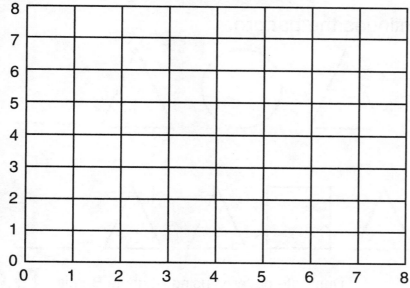

Name _____ Date _____

PATTERNS AND COORDINATE GRAPHS

Completing a Pattern

Ring the one that continues the pattern.

1.

2.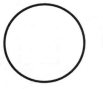

3.

4.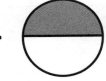

Draw the shape that continues the pattern.

5.   ___

6.   ___

Duplicate patterns using Attribute Beads.

www.svschoolsupply.com
© Steck-Vaughn Company

Unit 7: Patterns and Coordinate Graphs
Geometry 2, SV 5806-X

PATTERNS AND COORDINATE GRAPHS

Completing a Pattern

Draw the shape that continues the pattern.

1. △ ▢ ▢ △ ▢ ▢ △ ▢ ▢ △ (dashed)

2. ○ △ ▢ ○ △ ▢ ○ △ ▢ ___

3. ▢ ○ ▢ ▢ ○ ▢ ▢ ○ ▢ ___

4. △ △ ▢ △ △ ▢ △ △ ▢ ___

5. ▢ ▢ ○ ▢ ▢ ○ ▢ ▢ ○ ___

6. ▢ △ △ ▢ △ △ ▢ △ ___

Draw the shape that continues the pattern.

7. | L ⊔ ⊓ ⊓ L ⊔ ___

Duplicate patterns using Attribute Lacing Buttons. ▢

89

Name _____ Date _____

PATTERNS AND COORDINATE GRAPHS

Completing a Pattern

Draw the shape that continues the pattern.

1. ○ ○ △ ○ ○ △ ○ ○ △ ___

2. ▭ ○ △ ▭ ○ △ ▭ ○ △ ___

3. △ ○ ▫ △ ○ ▫ △ ○ ▫ ___

4. ▫ ▫ △ ▫ ▫ △ ▫ ▫ △ ___

5. ▭ ○ ○ ▭ ○ ○ ▭ ○ ○ ___

Draw the shape that continues the pattern.

6. | ⌊ ⊔ ▫ | ⌊ ⊔ ▫ ___

7. ╱ ⋀ △ ╱ ⋀ △ ___

Duplicate patterns using Sorting Beads.

www.svschoolsupply.com
© Steck-Vaughn Company

Unit 7: Patterns and Coordinate Graphs
Geometry 2, SV 5806-X

Name_____ Date _____

PATTERNS AND COORDINATE GRAPHS

Problem Solving

Look at the shaded part of the circle.
Do you see a pattern?

Ring the one that continues the pattern.

1.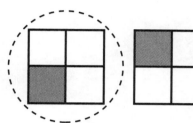

2.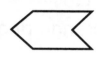

3.

4.

5.

PATTERNS AND COORDINATE GRAPHS

Using a Coordinate Graph

Follow these directions. Draw a ring at the crossing point.

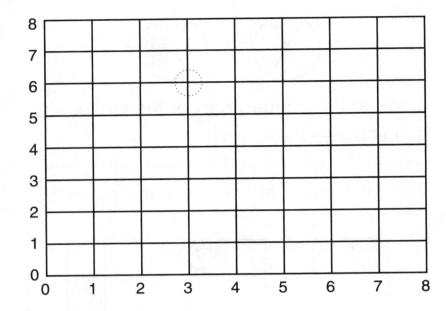

1. Start at 0. Go ⟶ 3. Go ↑ 6.
   Draw an orange ring at the crossing point.

2. Start at 0. Go ⟶ 5. Go ↑ 5.
   Draw a red ring at the crossing point.

3. Start at 0. Go ⟶ 7. Go ↑ 4.
   Draw a green ring at the crossing point.

4. Start at 0. Go ⟶ 6. Go ⟵ 6.
   Where are you? Ring the answer.

   0   3   6

Draw chalk coordinate graph on blackboard. Follow directions to walk to crossing points.

Name_____ Date _____

PATTERNS AND COORDINATE GRAPHS
Using a Coordinate Graph

The grid shows the county fairgrounds.

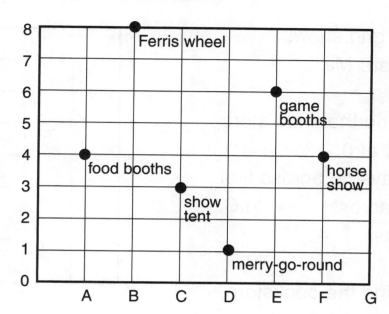

Ring the correct answer.

1. Go → to E. Go ↑ to 6.   Where are you?   Ferris wheel
   (game booths)

2. Go → to A. Go ↑ to 4.   Where are you?   show tent
   food booths

3. Go → to D. Go ↑ to 1.   Where are you?   merry-go-round
   show tent

4. Go → to B. Go ↑ to 8.   Where are you?   game booths
   Ferris wheel

Draw chalk coordinate graph on blackboard. Follow directions to walk to crossing points.

Name _____   Date _____

PATTERNS AND COORDINATE GRAPHS
Problem Solving

The grid shows Square Mall.

To find the shoe store, start at 0.
Always go across first.
Go across ⟶ to C.
Go up ↑ 4.

To find the bookstore, start at 0.
Go across ⟶ to G.
Go up ↑ 2.

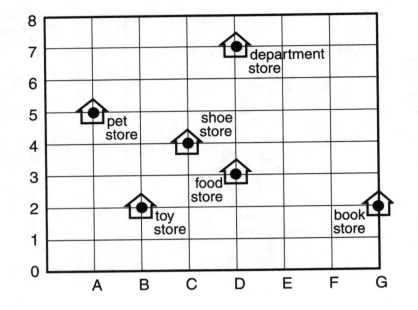

Ring the correct answer.

1. Go ⟶ to B. Go ↑ 2.
   Where are you? (toy store)   pet store

2. Go ⟶ to D. Go ↑ 7.
   Where are you?   food store   department store

3. Go ⟶ to A. Go ↑ 5.
   Where are you?   pet store   department store

4. Go ⟶ to D. Go ↑ 3.
   Where are you?   food store   shoe store

Draw chalk coordinate graph on blackboard. Follow directions to walk to crossing points.

www.svschoolsupply.com
© Steck-Vaughn Company

94

Unit 7: Patterns and Coordinate Graphs
Geometry 2, SV 5806-X

ANSWER KEY
Geometry for Primary Grade 2

UNIT 1: 2-Dimensional Figures
Assessment
P. 9
1. Student will correctly color the shapes.
2. 2 circles, 5 triangles, 4 rectangles
3. Student will correctly color the shapes.
4. 2 triangles, 3 circles, 5 squares
5. Student will draw a triangle.

P. 10
Student will ring each corner red and trace each side blue.
1. 4 sides, 4 corners
2. 0 sides, 0 corners
3. 3 sides, 3 corners
4. 4 sides, 4 corners
5. 4 sides, 4 corners
6. 3 sides, 3 corners
7. 6 corners

P. 11
1. 4 sides, 4 corners, 2 square corners
2. 6 sides, 6 corners, 0 square corners
3. 3 sides, 3 corners, 0 square corners
4. 4 sides, 4 corners, 0 square corners
5. 4 sides, 4 corners, 0 square corners
6. 5 sides, 5 corners, 2 square corners

P. 12
1. yes, 2. no, 3. yes, 4. no
5. 4 sides, 4 corners, 4 square corners
6. 5 sides, 5 corners, 2 square corners

P. 13
1. yes, 2. no, 3. no, 4. yes
5. 4 sides, 4 corners, 4 square corners
6. 4 sides, 4 corners, 4 square corners
7. 5 sides, 5 corners, 2 square corners
8. 4 sides, 4 corners, 2 square corners

P. 14
1. 4 sides, 3 corners, 1 square corner
2. 4 sides, 4 corners, 0 square corners
3. 4 sides, 4 corners, 2 square corners
4. 4 sides, 4 corners, 1 square corner
5. 3 sides, 3 corners, 1 square corner
6. 3 sides, 3 corners, 0 square corners

P. 15
1. Student will draw an octagon.
2. Student will draw a pentagon.
3. Student will draw a triangle.
4. Student will draw a square.
5. Pairs of students should draw the same shape.

P. 16
1. 3, 2. 4, 3. 0, 4. 4, 5. 4, 6. 4, 7. 0, 8. 4, 9. 3

P. 17
Two drawn shapes will vary throughout.
1. Ring second and third shapes
2. Ring first and third shapes
3. Ring second shape
4. Ring second shape

P. 18
5 circles, 4 squares, 4 triangles, 7 rectangles

P. 19
1. 2, 2. 4, 3. 3, 4. 6, 5. 13, 6. 14
7. rectangle, 8. circle

P. 20
1. Student will correctly color the shapes.
2. 2 circles, 5 triangles, 4 rectangles
3. Student will correctly color the shapes.
4. 2 triangles, 3 circles, 5 squares

P. 21
Student will correctly color the shapes.
1. 4 triangles, 3 circles, 4 rectangles
2. 7 squares, 11 triangles, 5 circles

P. 22
1. 0, 2. 2, 3. 0, 4. 4, 5. 3, 6. 4, 7. 0, 8. 4

P. 23
Students will accurately draw the shapes.

P. 24
1. Draw a square, 2. Draw a rectangle,
3. Draw a triangle, 4. Draw a hexagon,
5. The two shapes should match.

P. 25
1. Draw an octagon, 2. Draw a triangle, 3. Draw a square, 4. The two shapes should match.

P. 26
1.

Shape	Number
circle	2
square	5
rectangle	5
triangle	8

2.

birds	1	2	3	4
rectangles	5	10	15	20

3. 20, 4. 15

P. 27
1. 0, 0, 2. 3, 3, 3. 4, 4, 4. 4, 4, 5. 16 without the dotted lines, 44 with dotted lines

P. 28
1. Ring the second rug, 2. 1, 3. Ring the diamond, 4. Draw a pentagon

Unit 2: 3-Dimensional Figures
Assessment, P. 29
1. 3 cubes, 5 rectangular prisms, 4 cones, 2 cylinders, 2 spheres, 1 pyramid
2. Ring the square.
3. Ring the sphere.

P. 30
1. sphere, 2. rectangular prism, 3. cylinder, 4. cube, 5. cone

P. 31
1. cone, 2. cube, 3. cylinder, 4. Student will correctly match shapes.

P. 32
1. sphere, 2. cylinder, 3. cube, 4. cone
5. rectangular prism

P. 33
1. ice cream cone, megaphone
2. block, tissue box
3. can, jar
4. cracker box, brick
5. globe, second balloon
6. wheel, dime

P. 34
1. TV set
2. paint bucket, glass
3. party hat, construction cone
4. door, quilt
5. triangle, pennant
6. plate, record

P. 35
1. Student will accurately color the animals.
2. 5 spheres, 3 cubes, 8 cones, 5 cylinders, 7 rectangular prisms, 1 pyramid

P. 36
1. Student will accurately color shapes.
2. 3, 3. 3, 4. 1, 5. 2, 6. 1, 7. 2
3. Ring the third box.

P. 37
1. square, 2. circle, 3. rectangle,
4. circle 5. yes, 6. no, 7. yes, 8. no

P. 38
1. second circle
2. square
3. first circle
4. last rectangle
5. second circle
6. last rectangle
7. circle

P. 39
1. yes, 0, 0
2. yes, yes, 1, 1
3. yes, 2, 0
4. no, yes, 6, 8
5. no, yes, 6, 8

Unit 3: Congruence
Assessment, P. 40
1. Ring last figure
2. Ring third figure
3, 4. Students will accurately draw figures.

P. 41
1. Ring second figure.
2. Ring third figure.
3. Ring fourth figure.
4. Ring first figure.
5. Ring third figure.
6. Ring second figure.

P. 42
1. Ring second figure.
2. Ring fourth figure.
3. Ring third figure.
4-7. Students will accurately draw figures.

P. 43
1. Ring second figure.
2. Ring third figure.
3. Ring third figure.
4-7. Students will accurately draw figures.

P. 44
Student will accurately draw figures.

P. 45
Student will accurately draw figures.

P. 46
1, 2. Student will accurately draw patterns.
3. Color third figure.

Unit 4: Symmetry
Assessment, P. 47
1. no, 2. no, 3. yes
4-6. Answers may vary but should be lines of symmetry.
7-9. Answers may vary but should be two lines of symmetry.

P. 48
1. no, 2. yes, 3. yes, 4. yes, 5. no, 6. no
7-12. Answers may vary but should be lines of symmetry.

P. 49
1. yes, 2. no, 3. yes, 4. yes, 5. no, 6. no,
7. yes, 8. yes, 9. no
10-12. Answers may vary but should be lines of symmetry.

P. 50
1-9. Answers may vary but should be lines of symmetry.
10. Ring second and third figures

P. 51
1-5. Answers may vary but should be lines of symmetry.
6-11. Answers may vary but should be two lines of symmetry.

P. 52
1. Ring second and third figures
2. Answers may vary but should be lines of symmetry.

Unit 5: Perimeter and Area
Assessment, P. 53
1. 2 + 2 + 2 + 1 = 6 cm
2. 1 + 3 + 1 + 3 = 8 cm
3. 1 + 2 + 2 = 5 inches
4. 1 + 2 + 1 + 2 = 6 inches
5. 6 square centimeters, 6. 9 square centimeters

P. 54
1. 3 + 3 + 4 = 10 cm
2. 3 + 3 + 3 + 3 = 12 cm
3. 3 + 7 + 7 = 17 cm

P. 55
1. 4 + 3 + 6 = 13 centimeters
2. 8 + 4 + 8 + 4 = 24 centimeters
3. 4 + 2 + 4 + 3 = 13 centimeters

P. 56
1. 6 + 5 + 4 = 15 centimeters
2. 3 + 7 + 3 + 7 = 20 centimeters
3. 8 + 3 + 6 + 5 = 22 centimeters

ANSWER KEY
Geometry for Primary Grade 2

P. 57
1. $1 + 1 + 1 = 3$ in.
2. $2 + 2 + 2 + 2 = 8$ in.
3. $3 + 1 + 3 + 1 = 8$ in.
4. $3 + 3 + 1 = 7$ in.

P. 58
1. $2 + 4 + 4 + 2 = 12$ cm
2. $2 + 3 + 2 = 7$ cm
3. $3 + 3 + 3 + 3 = 12$ cm
4. $3 + 4 + 5 = 12$ cm
5. 6 square centimeters
6. 7 square centimeters

P. 59
1. $2 + 1 + 1 + 1 = 5$ inches
2. $1 + 2 + 2 = 5$ inches
3. $1 + 2 + 1 + 2 = 6$ inches
4. $1 + 1 + 1 + 1 = 4$ inches
5. 5 square inches
6. 3 square inches

P. 60
1. $1 + 3 + 1 + 3 = 8$ in.
2. $4 + 1 + 4 + 1 = 10$ in.
3. $1 + 1 + 1 = 3$ in.
4. 5 square inches

P. 61
1. 7 meters, 5 meters, 7 meters
2. 7 meters
3. 5 meters
4. $7 + 5 + 7 + 5$
5. 24 meters

P. 62
1, 2. Lauren's garden 2 meters x 6 meters,
Add 5 plants to picture.
Gary's garden 4 meters x 6 meters
Jo's garden 8 meters x 2 meters,
Add 7 plants to picture.
3. 16, 4. 16, 5. 20, 6. 20, 7. Gary's

P. 63
1. 500 feet, 2. 600 feet, 3. 700 feet, 4. 400 feet

P. 64
1. <400, 2. >500, 3. >500, 4. <900, 5. >800
6. >400, 7. >400, 8. <800

P. 65
1. 18 , 2. 17, 3. 22, 4. 20

P. 66
1. 4, 2. 3, 3. 6 , 4. 5

P. 67
1. 9, 2. 4, 3. 12, 4. 16, 5. 6, 6. 12

P. 68
Answers will vary for guesses.
1. 10 squares, 2. 6 squares, 3. 6 squares

Unit 6: Fractions Using Pictorial Models
Assessment, P. 69
1. 3, 2. 6, 3. 4, 4. $\frac{3}{5}$, 5. $\frac{2}{3}$, 6. $\frac{2}{6}$, 7. No

P. 70
1. 2, 2. 4, 3. 0, 4. 3, 5. 2, 6. 5, 7. 0, 8. 4, 9. 5

P. 71
1. 3, 2. 6, 3. 4, 4. 0, 5. 5, 6. 8
7-12. Student should draw lines to show equal parts.

P. 72
1. 2, 2. 4, 3. 8, 4. 6, 5. 2, 6. 8
7. Color the last figure.
8. Color the second figure.
9. Color the last figure.

P. 73
1. Color the third figure.
2. Color the first figure.
3. Color the second figure.
4. Student should draw lines to show equal parts.

P. 74
1. 4, 2. 3, 3. 2, 4. 4, 5. 6, 6. 6, 7. 2, 8. 0
9-11. Student should draw lines and write the number of equal parts.

P. 75
1. 8, 2. 3, 3. 4, 4. 6, 5. 2, 6. 2, 7. 10, 8. 8
9. Student should draw lines to show halves.

P. 76
1. 1, 2. 1, 3. 1, 4. 8, 5. 3, 6. 5, 7. 1, 8. 10, 9. 6
10-12. Student will shade one part of each figure.

P. 77
1. $\frac{3}{4}$, 2. $\frac{1}{2}$, 3. $\frac{1}{3}$, 4. $\frac{3}{5}$, 5. $\frac{5}{6}$, 6. $\frac{2}{3}$, 7. $\frac{4}{8}$, 8. 2, 9. 6

P. 78
1. 2, 2. 2, 3. 1, 4. 3, 5. 4, 6. 5, 7. $\frac{3}{4}$, 8. $\frac{2}{10}$
9. $\frac{2}{3}$, 10. $\frac{3}{5}$, 11. $\frac{6}{8}$, 12. $\frac{4}{10}$

P. 79
1. 1, 2. 1, 3. $\frac{1}{5}$, 4. 4, 5. $\frac{1}{5}$, 6. 3, 7. 1, 8. 8, 9. $\frac{1}{4}$
10. $\frac{1}{10}$, 11. $\frac{1}{2}$, 12. 10

P. 80
1. $\frac{2}{4}$, 2. $\frac{3}{8}$, 3. $\frac{4}{5}$, 4. $\frac{3}{8}$, 5. $\frac{2}{3}$, 6. $\frac{2}{6}$, 7. $\frac{5}{8}$, 8. $\frac{4}{10}$
9. $\frac{3}{3}$, 10. $\frac{3}{4}$, 11. $\frac{2}{5}$, 12. $\frac{2}{7}$

P. 81
1. 2, 2. 3, 3. 3, 4. 2, 5. 4, 6. 3, 7. $\frac{2}{5}$, 8. $\frac{2}{4}$
9. $\frac{2}{6}$, 10. $\frac{4}{6}$, 11. $\frac{5}{8}$, 12. $\frac{3}{5}$

P. 82
1. 1, 2. 2, 3. 5, 4. 5, 5. $\frac{2}{8}$, 6. $\frac{2}{4}$, 7. $\frac{4}{5}$, 8. $\frac{3}{8}$
9. Student should color 3 parts.

P. 83
1. Student will color $\frac{1}{2}$ of each figure
2. Student will color $\frac{1}{3}$ of each figure
3. Student will color $\frac{1}{4}$ of each figure
4. No

P. 84
1. Color $\frac{1}{3}$ of first, second, and fourth figures
2. Color $\frac{1}{4}$ of second and fourth figures
3. Color $\frac{1}{8}$ of second and fourth figures
4. Yes

P. 85
1. Color $\frac{1}{3}$ of first, third, and fourth figures
2. Color $\frac{1}{4}$ of first, second and fourth figures
3. Color $\frac{1}{8}$ of second and third figures
4. No

P. 86
1. Color whole board, 1
2. Color one half of board one color, other half another color, $\frac{1}{2}$, $\frac{1}{2}$
3. Color one third of board one color, $\frac{2}{3}$ another color, $\frac{1}{3}$, $\frac{2}{3}$
4. Color $\frac{1}{4}$ one color, 3/4 another color, $\frac{1}{4}$, $\frac{3}{4}$
5. $\frac{1}{4}$, $\frac{1}{3}$, $\frac{1}{2}$, $\frac{2}{3}$, $\frac{3}{4}$, 1

Unit 7: Patterns and Coordinate Graphs
Assessment, P. 87
1. rectangle, 2. triangle, 3. first figure, 4. (1,3)

P. 88
1. triangle, 2. square, 3. shaded triangle,
4. completely shaded circle, 5. circle, 6. square

P. 89
1. triangle, 2. circle, 3. square, 4. triangle,
5. rectangle, 6. triangle, 7. rectangle

P. 90
1. circle, 2. rectangle, 3. triangle, 4. square,
5. rectangle, 6. |, 7. /

P. 91
1. first figure, 2. second figure, 3. second figure,
4. first figure, 5. first figure

P. 92
1. (3,6), 2. (5,5), 3. (7,4), 4. 0

P. 93
1. game booths, 2. food booths,
3. merry-go-round, 4. Ferris wheel

P. 94
1. toy store, 2. department store, 3. pet store,
4. food store